STEAM FOR THE MILLION.

A

POPULAR TREATISE

ON STEAM,

AND ITS

APPLICATION TO THE USEFUL ARTS,

ESPECIALLY

TO NAVIGATION.

INTENDED AS

AN INSTRUCTOR FOR YOUNG SEAMEN, MECHANICS' APPRENTICES
ACADEMIC STUDENTS, PASSENGERS IN MAIL STEAMERS, ETC.

BY

J. H. WARD.

COMMANDER, U.S.N.,
AUTHOR OF "NAVAL TACTICS," ETC., ETC.

A NEW AND REVISED EDITION.

Fredonia Books
Amsterdam, The Netherlands

A Popular treatise On Steam and its Application to the Useful Arts, Especially to Navigation

by
J. H. Ward

ISBN 1-58963-047-5

Copyright © 2001 by Fredonia Books

Reprinted from the 1867 edition

Fredonia Books
Amsterdam, The Netherlands
http://www.FredoniaBooks.com

All rights reserved, including the right to reproduce this book, or portions thereof, in any form.
In order to make original editions of historical works available to scholars at an economical price, this facsimile of the original edition of 1867 is reproduced from the best available copy and has been digitally enhanced to improve legibility, but the text remains unaltered to retain historical authenticity.

PREFACE.

THE author's present purpose is, to strip from the subject of steam the veil of mystery which hides it from popular view; a purpose, best accomplished perhaps by a mind more used in the stirring world of practice, than in the study, the engine-room, or work-shop. Or to change the figure, and assume the propriety in our age and country of bridging the chasm which separates the professions from the million, it is proposed, in this instance, to throw the span from the popular side; and the author would propitiate those belonging to the domain entered, by expressing a hope, that they will regard the act neither as intrusive nor in any way unwarrantable.

The excellent popular articles appearing in the newspapers,—like, for example, those which have described and discussed such marvels of art as the "Adriatic" and "Great Eastern,"—fail in the effect justly due to their superior merit, simply because the general mind is not systematically instructed by an elementary book free from all technicality of expression or intricacy of calculation. If the present one shall, by its plan, which is sufficiently apparent from the table of contents, and by its style of execution, so far open up the subject as to produce a nicer and more general appreciation of those and kindred articles, no small good will have been effected.

Passengers on ocean-steamers, suffering from ennui and impatience, may find in these pages pleasant relief. By engaging attention, they will beguile the time and thus shorten a passage; for as proverbially " a watched pot is longest boiling," so is the passage longest which is watched day by day. Abstraction is a happy faculty; and the employment which induces it, next to religion the best solace. Obviously at sea, no other occupation is so useful or appropriate as a study of the grand and powerful motor which hurries one over the waves; especially considering that the boilers and engines, in constant operation, exist as if for the very purpose of study, and beckon to it with each revolution of the crank. Such an opportunity at hand for acquiring this useful knowledge, easily and without detriment to other pursuits, ought not to be neglected.

It remains for the author to express his acknowledgments to the Engineers of the Navy, unsurpassed as a body in professional accomplishment, for the cheerful courtesy with which information sought from them is always rendered; and this recognition is most especially due to the author's friend, Chief Engineer J. W. KING, U. S. N., for an offer of the manuscript pages of his forthcoming work to be entitled " Lessons for Young Marine Engineers:" also to the finished draftsman and mechanic of his office, Mr. JOHN V. VAN DOREN.

CONTENTS.

Section I.

Introduction, 7. Source of power, 9. Nature of steam, 11. Evaporation and condensation, 11. Ebullition, and formation of steam, 13. The atmosphere the unit of pressure, 15. Methods of determining pressures, 16. Foaming, the cause and remedy, 20. Reduction of temperature without a reduction of pressure, 21. Effect of ebullition on water level, 22. Preservation of water level, 23. Maintaining the circulation, 24. Saturated and surcharged or superheated steam, 25. Blowing sediment, or saturated water, 26. Volume of steam—computing evaporation, 29.

Section II.

Theory of latent and sensible caloric, as applied to steam. Specific Heat, 35.

Section III.

Construction of boilers—fire surface and flues, 36. Flue boilers, 39. Tubular boilers, 40. Martin's vertical water tube boiler, 40. Water room and steam room, 42. Strength of boilers and metal of which they are made, 44.

Section IV.

Combustion, 46. Draft, 47. Blast, 47. Fuel, 48. Smoke, 50.

Section V.

The steam engine, 52. Interior construction of the cylinder and condensing apparatus, 52. General description of the engine, 56. High and low pressure, 63. Surface condensation, 64. Heaters for feed water, 66. Expansive action of steam, 67.

CONTENTS.

SECTION VI.

The high pressure engine, 73. Relative advantage of the high and low pressure engines, 73. Locomotive engines, 76. Calculated and indicated horse power, 77. Consumption of fuel and Marine Economy, 82. Raising steam and managing it and the engine, 86. Distillation, 90. Link Motion, 91.

SECTION VII.

Construction of steamers, 94. Side wheels and *screws, 100. Wheel shafts, 105. The screw shaft, 106. Coupling and lifting screws, 109. Elements of the screw, 112. †Steerage, 116. Conclusion, 118.

---

* Want of room in the note page 108 prevented further comment on fouling screws, called for by its frequency. The "Exmouth" ship of the line obstructed her screw by winding upon it her own sheet cable laid out to haul off a shoal. The "Melbourne" screw packet was dismasted in a gale, the rigging fouled and disabled her screw, and thus doubly disabled she laid the gale out, helpless like a log. These instances are mentioned, with many others, by Sir Howard Douglas in his "Naval Warfare with Steam," p. 63 *et seq.*, where several remedies are proposed, and a very proper precaution, viz., providing every screw-vessel with a suit of submarine armor. To men-of-war in fleets this is peculiarly important.

The "Princeton," the "San Jacinto," and many others in our Navy have had a hawser wound on the screw; and very recently the same happened to the Niagara in New York harbor. The latter case is described in detail by the following extract. The hawser was a Manilla, which had been paid out as a tow line to a tug boat, and when cast off by the tug, was not gathered in, but left, about 20 fathoms of it, hanging from the stern.

"The hawser caught across one of the blades, and after being wound six times around the hub abaft the blades, was forced down to the journal, where it jammed between the shoulder of the journal and the pillow block of the frame; and it jammed there so tightly that no more could wind on, but the journal revolved within the six turns."

The hawser was thus forced into the narrow space described, and no more wound on to the screw, because the end was fast on deck, and when the revolutions had wound up all the slack, either it was to part or tighten on to the journal, and jam as it did.

The crank brasses subsequently heated, causing the vessel to anchor. This is fairly attributable, in part at least, to obstruction of the screw by the hawser.

To raise the screw out of water for clearing the hawser, a force was required three times that necessary to overcome its weight, for the hawser as jammed had sprung open the screw frame, causing it to bind between the stern posts.

† The habit, beginning to prevail in men-of-war, though not perhaps very greatly objectionable in merchant-vessels, that of running ships off the wind or off the course upon trifling occasions, is one which will cost severe disaster by collisions when we come to sail in fleets, as we one day must. The habit should not grow. As a practice, it was formerly unknown, and is rarely necessary in ships well manned with well-trained crews.

STEAM.

SECTION I.

Introduction. Source of the power exhibited by steam engines. Nature of steam. Evaporation and Condensation. Ebullition of water, and formation of steam under different pressures. The atmosphere the unit of measurement in determining pressure. Pressures in a boiler, and methods of determining them. Foaming in boilers,—the cause and remedy. Effects of a reduction of temperature in the water of a boiler, without a corresponding reduction of pressure. Effects of ebullition on the level of water. Preservation of water level. Maintenance of circulation. Saturated and surcharged or superheated steam. Blowing off sediment and saturated water.

INTRODUCTION.

THE Author, with a keen sense of the obstacles which seamen encounter in acquiring knowledge of steam, recently applied on the ocean, undertakes with his experience to aid younger men of the nautical profession in overcoming such hindrances. This aid, he as a seaman may render, more effectually perhaps than an engineer, who has not gained knowledge in the same irregular way by which we must attain ours, and may therefore be presumed to possess but an imperfect idea of the difficulties and obscurities peculiar to our path, and of the simpler and clearer language and arrangement needed to smooth and enlighten it, and facilitate progress.

Whilst a youth preparing for nautical command should, above all things, steadily adhere to the purpose of becoming a thorough seaman—familiar with the ocean in all its aspects, the winds and the waves, and the means of averting their sometimes awful force, the new motor now more and more common, demands no inconsiderable share of his attention. Every such person is liable, in the course of service, to be thrown, in positions of responsibility, on board ships driven wholly or in part by steam; and if wise, he will not heedlessly await the event, but qualify himself for it, by gathering such information as will establish an intelligent concert between himself and the engineers with whom he may be associated.

Yet it is not intended to inculcate the idea that the seaman should aim to accomplish himself as an engineer, but the contrary; for no man can become an adept at two professions so distinct, and each so comprehensive. Thus, whilst the competent marine engineer, besides a good preliminary education and skill as a practical workman and draftsman, needs ample knowledge in construction, in mechanical detail, and in strength as well as fitness of materials; to have his mind stored with facts for data, and formularies for calculation; to know the qualities of fuel, and how to economise its consumption; also, by long use to have acquired confidence and sleight in handling engines of various forms;—the seaman, requires only to be ordinarily grounded in general principles; to possess such an outline knowledge of construction and operation as will enable him, when in command, to comprehend, and intelligently to judge reports, explanations, recommendations, and the practice of engineers; to ascertain, generally by silent observation, as well what an engineer as any other officer is about, and consequently to know when, but more especially when *not*, to interfere in the pre-eminently important department, immediately and specially controlled by the chief engineer of a ship.

This interference ought to, and will of course, considering the high character usual in those selected for that important post, rarely be called for. But when it does become necessary, the commander should be capable of rendering it prompt and effectual, yet not irritating, nor in a manner to impair the engineer's interest, or relieve his responsibilities. In fine, the seaman's knowledge of steam needs to be such as fits him for administrative, rather than for mechanical duties, scientific or practical.

The Persian maxim, " whilst ignorant, let not shame prevent thy asking questions," is excellent, provided the querist possesses knowledge enough to give point and direction to his inquiries, and elicit patient replies; for few people relish an interrogation, which carries evidence of total ignorance in the first principles of its object. The instruction here given will suffice to guide such inquiry, by which every boiler and engine may become a subject of interest and source of instruction, directly itself or through the person in charge, and knowledge constantly and imperceptibly grow out of experience attending the daily walk and occupation.

SOURCE OF POWER.

1. Mechanical *prime-movers*, may be classed under two heads—those which act by elastic force, as steam, the gases developed by inflamed gunpowder, heated air, and steel springs; and those which act by the force of gravity, as suspended weights, atmospheric air, and water-falls.

2. The force exerted by these agents, in no case, except perhaps that of gunpowder and of the winds, acts immediately upon the ultimate object, but mediately through mechanism—as the steam-engine, the system of wheels and balance or regulating pendulum in time-keepers, and the water-wheel of a mill.

3. The latter is the simplest, and the most familiar application of power to its object through intervening machinery. The power exerted depends upon the head, and upon the quantity of water which is discharged on, or against, the water-wheel. And that wheel is considered the best, whether it be an overshot, an undershot, or a tub-wheel, which uses the water most economically—with least waste. The wheel creates no power, but derives its efficacy solely from the head and quantity of water.

4. So in the case of a steam-engine; the power exerted depends upon the elastic or expansive force of steam in the boiler, also upon the quantity of steam the boiler can supply to work the engine and yet maintain the given or required elasticity. And in general, the best engines are those which use the steam most economically—with least waste through leakage, condensation, or friction—in conveying the power to any desired end, as raising water, driving mills, or propelling vessels.

5. Yet the engine, though creating and having none in itself, is commonly spoken of as possessing power. But this is only a convenient figure of speech, expressive of the capacity and strength of parts to receive and transmit a specified amount of power. Evidently, if a cylinder lacks capacity and strength to contain steam sufficient in quantity and elasticity to exert a given power, or if the other parts of

EVAPORATION AND CONDENSATION. 11

the machinery lack strength to convey the power which presses upon the piston, the engine with parts thus lacking cannot be said to be of that given power.

6. Steam therefore, being the source of power as it certainly is of danger, a discussion of its nature, the manner of its production, and the various phenomena attending its existence, are properly first in order.

NATURE OF STEAM.

7. Steam is the vapor of water, and results from a certain quantity of heat entering into combination with water. After steam is formed, applying to it an undue amount of heat does not materially increase its pressure, because there is no additional quantity of water for the increase of heat to enter into combination with.* But supply water along with the heat, and the two, entering into combination, produce an increase of steam. On the other hand, abstract the heat, and the steam becomes condensed,—reduced again to water—wholly or in part, just in proportion as the heat is abstracted.

EVAPORATION AND CONDENSATION.

8. The surface of water exposed to the air always gives a spontaneous vapor, which is visible or invisible according to the relative temperature of the water and the air. If the water be warm and air cool, as in autumnal mornings, the vapor, which has equal temperature with the water, is condensed in the cooler air and becomes visible in fog.

---

* This is true of common dry or super-heated steam, but not of that raised by excessive heat to the gaseous state, when it is termed *stame*.

9. But all vapor, whether it be the result of spontaneous evaporation, or steam of the highest temperature produced in a boiler, is transparent unless condensed by exposure to a cooler temperature. If a boiler were of glass, nothing within would be visible but water. But lift the valve, and let steam escape into cooler air, and a cloud instantly appears, which though termed steam, is more properly condensed steam, through loss of its heat; and to the extent it is so condensed it has lost its force.

10. The breath is an illustration. It is always loaded with moisture when exhaled. The moisture is usually visible, however, only upon cool mornings, because the temperature is then low enough to produce condensation. Breathe upon a cooler solid, and the moisture, by condensation, becomes visible in a liquid form.

11. The same thing is manifest upon the surface of a pitcher of ice water which condenses the moisture existing in the natural atmosphere; in the fog visible about an iceberg at sea; in the dew which forms at night upon the earth; and is that which makes the warm winds from the ocean or from southern latitudes, damp winds.

12. But the vapors produced by spontaneous causes and floating in the air, possess none of the mechanical force of steam as used in the arts. This force in steam results only from an application to water, of high heat produced by artificial means; and the quantity and force of the steam formed, will depend upon the intensity of the heat, and the extent of fire or heating surface to which it is applied—for although steam is given off at the "water surface," its formation takes place at, and is due to the "fire surface."

EBULLITION OF WATER, AND FORMATION OF STEAM, UNDER DIFFERENT PRESSURES.

13. Fire surface in a boiler, is that surface of metal boiler plate, which is surrounded by, or is in contact with, and receives protection from water, and is acted upon by heat sufficiently high in temperature to produce steam.

14. Solids, convey heat by their conducting powers. Thus the interior surface of a flue or furnace becomes heated; this heat is conducted to the exterior surface, and from that passes into the water which lies against it.

15. But heat is not distributed throughout a body of water by its conducting power, for water is not a conductor of heat. It is disseminated in water by circulation of the particles, which brings all of them successively in contact with the heated metal, or fire surface.

16. The currents produced by this circulation are principally in vertical planes, the heat passing upwards from the fire surface. Hence, water boils above the level of the fire surface, but not below it. This is illustrated by putting a lamp to a glass tube filled

with water. From b upwards, the water will boil. Below b, it will not boil. So the lower parts, or water legs of a boiler, do not give signs of much heat in the water they contain.

17. The theory of ebullition is as follows: The stratum of water next the fire surface becomes heated and rarefied; being lighter than the water above, this heated stratum rises to the surface; there it gives off a portion of its heat in vapor, becomes in consequence cooled and of greater specific gravity and sinks; again it receives heat from a contact with the fire surface and again rises; by which process of circulation, every stratum of water comes successively in contact with the heated metal surface, and the whole body of water becomes heated. If the heat of the fires be so great, or the fire surface so extensive, that heat is communicated to the water faster at the fire surface than it is given off in vapor at the water surface, an accumulation of heat will take place in the water until it reaches the temperature of 212 degrees, when the circulation will have become so rapid as to produce ebullition. When this ebullition takes place, heat escapes at the water surface in steam (if that surface be exposed to atmospheric pressure, the barometer standing at 30 inches) as rapidly as it can be communicated at the fire surface, and the temperature cannot be raised higher than 212°. Hence 212° is the boiling point of fresh water when the barometer stands at 30 inches. Sea water boils at 213°.

18. Vapor, being heat in combination with water, it follows that all evaporation absorbs and carries off heat in large quantities, reducing temperature rapidly.

19. Under a *decreased* atmospheric pressure upon the surface of water, produced by exhausting a receiver or by moving to a highly elevated position, water will boil at a temperature which reduces with the pressure; until when, as in a vacuum, where the

pressure is removed altogether, ebullition takes place at the temperature of 88°. A vessel of water heated to 88°, under an exhausted receiver will render the fact manifest. At the summit of Mont Blanc, pressure is so far decreased that water boils at 187°; beyond which temperature it cannot there whilst exposed to the air, be heated.

20. Under an *increased* pressure beyond that of the atmosphere, produced by confining and accumulating steam as it forms, the temperature at which water boils increases;—*the temperature of the steam formed, is equal to the temperature of the water from which it forms;—and the elastic force of the steam formed, is equal to the pressure under which it forms.* For example, under one atmosphere water boils at 212°; the steam formed is of that temperature, and has an elastic force equal to an atmosphere. Under two atmospheres, water boils at 250° of temperature, and produces steam of 250° temperature, and of an elastic force equal to two atmospheres, and so on.

THE ATMOSPHERE, THE UNIT OF MEASUREMENT IN DETERMINING PRESSURES.

21. The pressure of the atmosphere at the level of the ocean with the barometer at 30 inches, is taken as the unit in estimating and comparing pressures and elastic forces.

22. This atmosphere is an invisible fluid which surrounds the earth, resting upon it, and upon all things belonging to it, with a weight equal, at the level of the sea, to 14.7 pounds on every square inch of surface. Persons are not, however, aware of this weight by their sensations, because they experience an equilibrium of pressure—above, below, within and

without alike. And so of inanimate objects when an attempt is made to move them. Ordinarily they are pressed alike on all sides, and the atmosphere therefore opposes no effective resistance to their motion. But exhaust air from one side of any thing, the equilibrium is destroyed, and the amount of pressure immediately becomes manifest by its effects.

23. The question is, how was this pressure of the atmosphere ascertained to be 14.7 pounds upon the square inch? The column of atmosphere 45 miles high, and one square inch of area, just balances, and consequently weighs the same as a column of mercury of like area and 30 inches high. This column of air also balances 33.86 feet of water. Consequently a column of air, 30 inches of mercury, and 33.86 feet of water, weigh the same; and these two last weighing 14.7 pounds each, it follows that the weight or pressure of the atmosphere is equal to 14.7 pounds upon every square inch.

PRESSURES IN A BOILER, AND METHOD OF DETERMINING THEM.

24. Steam at 212°, having a pressure only equal to that of the atmosphere, just balances it, and can exert no mechanical effect in overcoming its resistance. An atmosphere of steam, therefore, in a boiler, takes the place of air that existed there before steam formed, and consequently just balances the atmospheric pressure without. Increase the steam to 250° of temperature, and the corresponding pressure of two atmospheres, and the internal will *exceed* the external pressure by an atmosphere, and so on.

25. All modes of measuring the pressure in a boiler, (other than by ascertaining the temperature) determine

STEAM GAUGE.   17

merely this *excess* of pressure (Art. 24,) beyond that of the external atmosphere.* These modes of measurement are, usually, the safety valve and the mercurial gauge.

26. The safety valve is a circular plate of a determined area, fitting in an aperture of the boiler, and loaded with a weight proportioned to the pressure of steam to be measured and to the area of the valve. If, for example, the steam which can safely be carried equals an *excess* of one atmosphere, (Art. 23,) or about 15 pounds per square inch, and the valve have an area of 10 inches, then a weight at all exceeding 150 pounds, placed on the valve, will keep it down in its seat, if the pressure of steam is not (in excess) more than an atmosphere. If the steam increase beyond that point, the valve will rise and steam escape. And when up, although it did not rise until the pressure had increased above 15 pounds, the valve will not fall again until the pressure is reduced considerably below 15 pounds. This is chiefly owing to the conical form of the valve—thus:

When the valve is down in its seat, the area on which steam presses to raise it, is calculated from the diameter cd; but when the valve is up, the steam acts upon an area calculated from the greater diameter ab.

27. This is usually the method practised to determine pressure in high pressure boilers, which carry from 3 to 20 atmospheres. The mercurial gauge (Art. 28,) is not used on those boilers, because the

* In speaking of pressure *this excess* is understood.

great pressure of steam would blow the mercury from the gauge tubes, unless of an inconvenient length.

28. The steam carried by low pressure boilers, (from 3 to 25 pounds), is measured by the mercurial gauge. This is a bent tube, or syphon inverted, having a long and a short leg, and partly filled with mercury. The orifice of the short leg fits, with a tight joint, to an opening in the boiler above the water level, and the long leg is open to air. The mercury, when not pressed upon by steam, may rise in the short leg nearly to the orifice at the boiler, and will of course stand at the same level in the other leg. And when an atmosphere of steam has taken the place of the atmospheric air within the boiler, the mercury will still continue at the same level in the two legs. But as steam increases, it will press with greater force on the mercury in the leg exposed to *steam*, than the atmosphere presses on the mercury in the leg open to *air;* and consequently the mercury will rise in the open leg, and indicate the excess of pressure in the steam beyond that of the external air.

The graduations on the mercurial gauge are an inch in length. Fifteen pounds of pressure per square inch equals the weight of nearly 30 inches of mercury, (Art. 23),—or one pound of pressure equals the weight of two inches of mercury. Every inch that the mercury rises in the gauge indicates an alteration of two inches in the level of the column supported in the bent tube. Consequently, one inch of mercury on the scale, indicates a pound pressure of steam.* And hence, inches of steam, and pounds of steam, are used in common language as convertible terms.

---

* "Strictly, a column of mercury 2 inches in height, will counterbalance a pressure of .98 lbs. on a square inch."—*Haswell.*

ELASTICITY AND TEMPERATURE.   19

Thus, let d be a boiler, on the face of which is set a gauge as described. Into this tube pour mercury

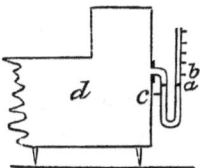

until it rises in both legs to the level a, near the orifice in the boiler. When steam is raised, and the pressure in the boiler exceeds that of the atmosphere by one pound, the mercury in the steam leg of the gauge will descend one inch to c, and consequently will rise one inch in the open leg to b, producing a difference in the levels c and b of two inches, which is the column sustained by one pound of steam.

29. An indirect method of determining, not the excess, but the absolute pressure of steam in a boiler, is to obtain its temperature, either by a thermometric instrument, or by the melting point of alloys, which are known to liquify at certain ascertained degrees of heat. If the temperature of steam can by either of these means be obtained accurately, and the steam be saturated (Art. 34), the corresponding pressures are known with certainty.

Thus a temperature of 212°* indicates a pressure of 1 atmosphere.
"   "   242   "   " $1\tfrac{1}{2}$ "
"   "   252   "   "   2   "
"   "   264   "   " $2\tfrac{1}{2}$ "
"   "   277   "   "   3   "
"   "   285   "   " $3\tfrac{1}{2}$ "
"   "   294   "   "   4   "
"   "   301   "   " $4\tfrac{1}{2}$ "
"   "   308   "   "   5   "
"   "   314   "   " $5\tfrac{1}{2}$ "
"   "   320   "   "   6   "
"   "   326   "   " $6\tfrac{1}{2}$ "

* This is with fresh water. Sea water boils at 213—and supersalted water, at a still higher temperature.

FOAMING IN BOILERS, THE CAUSE AND REMEDY.

30. *Steam* does not form from water, until the latter has reached the point of moderate ebullition. And the temperature of the water so boiling, is similar to that of the steam which forms from it (Art. 20). Therefore, by Art. 29, water under one atmosphere of pressure boils at 212°; under 1½ atmospheres at 242°; and 2 atmospheres 250°; under 2½ atmospheres 264°; and so on, according to the table.

31. From this it is apparent, that in nature, a certain established relation exists between pressure upon the surface of water, and the temperature under which it moderately boils and forms steam. If the temperature of the water exceeds this relation, a violent ebullition or *foaming* will take place, and will continue, with a violence proportioned to the excess, until the natural relation is restored. If for example, water, under an atmosphere and a half of pressure of steam in a boiler be at the temperature of 242° (Arts. 29 and 30), and suddenly, by working steam off much more rapidly than it forms, the pressure is reduced to one atmosphere, the temperature of the water remaining at 242°, will exceed the established natural relation to pressure, and in consequence a violent ebullition or foaming will ensue. This operation may be illustrated thus: Heat a vessel of water to 212° in the open air, and the water will boil moderately. Place this vessel of water at 212° under the receiver of an air pump, and by partially exhausting the air, reduce the pressure a fraction only; the effect will be, by disturbing the natural relation between temperature and pressure, violent foaming in the water. Reproduce the pressure by confining the vapor formed, or by readmitting

air to the receiver, and the foaming will cease. So *in a boiler* which foams, close the throttle valve (Art. 7, Sec. 5) so as to reproduce the pressure of steam on the surface; then, the reduction of temperature by evaporation (Art. 18), and increase of pressure, both going on and conspiring, will soon restore the natural relation between the temperature of the water and pressure on its surface, and foaming will cease. This sudden reduction of pressure, and consequent foaming, are most likely to occur in boilers that have small steam room (k Art. 7, Sec. 3) as compared with the capacity of cylinder;—for, to take an extreme case, suppose a cylinder to consume 24 cubic feet of steam at each stroke, and there exists but three times that quantity of steam formed in the boiler, the consequence must be, at each stroke of the piston, a sudden reduction of $\frac{1}{3}$ in pressure; and a violent foaming will continually go on, owing to this drain, and to the rapid generation of steam necessary for the supply of succeeding strokes. Hence the rule, in constructing boilers, to give a large proportion of steam room as compared with the capacity of the cylinder.

EFFECTS OF A REDUCTION OF TEMPERATURE IN WATER, WITHOUT A CORRESPONDING REDUCTION OF PRESSURE.

32. If the temperature of water, and the pressure, stand both in a boiler at their natural relations; and suddenly a supply of cooler feed water be injected, in quantities so great in proportion to the water contained in the boiler as to sensibly reduce its temperature, the pressure remaining the same, the temperature of the water becomes reduced *below* the natural relation between temperature and pressure, and in consequence, the water ceases to boil or to make

steam, and will lie, with a surface as smooth and quiescent as that of molten lead, until the fires have raised the temperature, and working off steam has reduced the pressure, and restored the natural relation. Hence, the disadvantage of too great a reduction of water room in boilers, is guarded against in their construction (Art. 8, Sec. 3).

EFFECTS OF EBULLITION ON THE WATER LEVEL.

33. When steam in a boiler forms under moderate ebullition, and is worked off by the engine in quantities nearly equal to the supply, the surface of water, in consequence of the uniform ebullition, is raised, and stands constantly somewhat higher than the natural undisturbed level which would be taken by the same quantity of water not boiling. Hence if the engine stops, and steam accumulates so as to increase pressure beyond the natural relation (Arts. 31 and 32), the water will cease to boil, and its surface become quiescent at a level below that preserved whilst boiling. And therefore, although there may have been apparently a sufficient supply of water in the boiler when the engine was in motion, the water may fall when it stops, and expose the flues.* To prevent this, careful engineers, when the engine stops temporarily, let off steam by the safety valve nearly as fast as it had been worked off by the engine. A neglect of this precaution has undoubtedly often caused disasters; and a general attention to it may be taken as a tolerable indication of the carefulness of the engineer.

* This is illustrated in domestic life. A kettle of water boils over. Take it off the fire, it ceases to boil and the level falls. So in a boiler, if pressure upon the surface of water be increased, it ceases to boil and the level falls.

PRESERVATION OF WATER LEVEL.

34. Whenever from any cause, whether the feed or supply pumps becoming choked, neglect to turn on the supply, subsidence of water from cessation of ebullition, or otherwise, the water falls below the tops of the flues, and they thus become deprived of protection from water, they of course heat; and when water rises again upon this heated metal, steam forms suddenly, of an elasticity, and in quantities, greater than the boiler can bear, or than can find vent by so small an opening as the safety valve seat. Explosion is the natural and inevitable result. This is the most common cause of that disaster, many suppose the only one; and it is not unusual in investigations, to find all testimony rejected which does not conform to this theory.

35. To know the level of water in a boiler, and enable the engineer to judge when feed water is necessary to prevent exposure of the flues (Art. 34), boilers are provided generally with four gauge cocks; the highest cock is at a point above which the water cannot be permitted to rise without encroaching upon the room provided as a reservoir for steam;—and the lowest cock is at a point below which the water cannot be permitted to fall, without endangering exposure of the flues. The two intermediate cocks are convenient to show intermediate levels. If, when any one of these cocks is turned, it "blows" steam, the water of course is below the level of that cock;—and if it blows "solid water" (meaning, not foam), then the water is above that level. These cocks are generally placed on the front of every boiler.*

36. In addition to the gauge cocks, some boilers

---

* An experienced ear will detect water, foam, or steam, by the sound.

have also glass gauges for indicating the height of water. The glass gauge is a tube of that material, well annealed, so as not to break easily. This tube is connected with the boiler by two metal pipes; one pipe from the upper extremity of the tube connecting above the proper level of the water, and the other from the lower extremity connecting below that level. The water stands at the same level in the boiler and in the tube; consequently a glance of the eye at the tube shows where the level of water is in the boiler. These gauges stand very well—but sudden changes of temperature break them.

MAINTAINING THE CIRCULATION.

37. As equalization of heat throughout a body of water in a boiler is produced by circulation, which depends on evaporation (Arts. 15, 17, and 18), if evaporation is prevented by an undue accumulation of pressure, occasioned by stopping the engine, the inference is that circulation will cease, hence also that the equalization of heat will not go on, and that the stratum of water next the fire surface will become superheated, generating steam, at the fire surface, of a pressure corresponding with the local temperature of the water there: perhaps also forming globules of steam which may press the water away from the metal, and deprive it of the necessary protection from the water. In this state of things, if pressure be suddenly removed from the surface by starting the engine, the globules of steam thus released will rush to the surface and give off steam, in quantities, and of a pressure, greater than the boiler can withstand, and explosion will result.

This theory has been advanced. It is plausible,

and if true, constitutes an additional reason, when an engine stops, for opening the valve in order to maintain the circulation; and it accounts for many explosions, regarded as inexplicable on the generally received hypothesis, that although a weak spot in a boiler may give out under steady pressure, explosions never occur when there is plenty of water in the boiler.

SATURATED AND SURCHARGED OR SUPERHEATED STEAM.

38. When the temperature which steam possesses has been derived solely through the medium of the water, the heat has brought with it from the water a due proportion of moisture—all it can contain; and the steam so existing is termed *saturated steam*. But, if heat from any other source be taken up by steam, it becomes charged with a temperature disproportioned to its moisture, and is then termed *superheated* or *surcharged steam*. *Surcharged steam*, then, is steam heated to a temperature higher than is due to its pressure as represented in Art. 29. The opinion formerly prevailed, that water thrown into steam highly surcharged, would suddenly be converted to highly elastic steam and endanger explosion. Elaborate experiments made by the Franklin Institute of Philadelphia in 1836, show, however, with apparent conclusiveness, that no such effect results; but that on the contrary a small reduction of elastic force is produced by the injection of water to surcharged steam.

39. If saturated steam loses any of its heat by radiation from the steam pipes during its passage through them from the boiler to the cylinder, it condenses. But surcharged steam may radiate its superlative heat and not condense. Hence it is common to construct a steam drum around the base of

the chimney, by which the steam becomes superheated, at the same time that it affords some protection to the metal of the chimney against burning.

Another reason for heating the steam by the chimney is, that along with steam, some water, in the form of spray (called priming), is apt to rise from the surface, in which case by taking heat from the chimney it becomes steam. And all the heat that can be in any way turned to account, is so much saved from utter waste in the air.

BLOWING OFF SEDIMENT.

40. Feed water from fresh water rivers, holds vegetable and earthy matter in mechanical suspension. These, unless the water in boilers be occasionally changed, will deposit on the furnace and flues a crust which, being a non-conductor, will not only obstruct the communication of heat to the water, but will deprive the flues of the necessary protection from water. Waste of fuel is also a consequence, because if the heat cannot pass into the water, it goes off up chimney. To get rid, therefore, of the water containing the sediment before it deposits, boilers are provided with "blow pipes," leading generally from near the bottom of the boiler. When the "blow cock" is turned, the pressure of steam on the surface forces water through the blow pipe with great violence. It is this operation of blowing which is being performed, when a cloud of steam is seen under the wheel guards, accompanied with great noise, on board passage boats.

BLOWING OFF SATURATED WATER.

41. Boilers fed by salt water require a change, by blowing, very often. When evaporation takes place

from a body of salt water in a boiler, no salt is carried off in the steam;—that is wholly fresh. Manifestly, then, by constantly admitting salt to a boiler with the feed water, and passing none off in steam, the accumulation there of salt will continue, and will eventually fill the boiler with that substance unless means be adopted to get rid of it. For this purpose, therefore, in marine boilers, the "blow pipe" leads through the vessel's side or bottom, from about the crown of the furnace, where the saltest water is supposed to exist.

42. By blowing off highly saline water in this pipe, the excess of salt is removed; but a considerable quantity of heat also passes off with it, and is lost. Then to preserve the level of water in the boilers, that blown off must be replaced by other water thrown in by the feed pumps;—which water will require fuel to heat it. To avoid, therefore, any *useless* loss of heat and waste of fuel by blowing, care is observed to perform this operation no oftener, and to no greater extent, than is necessary to prevent the actual deposit of salt, or "scale" of some kind.

43. To determine when the necessity for blowing exists, careful engineers draw off occasionally from the boiler a small portion of water, and try its specific gravity by the hydrometer—for the salter the water, the greater its density or specific gravity.

Water holding all the salt possible in solution, termed a "saturated solution," more familiarly "saturated or super-salted water," when on the eve of depositing solid salt, is known to contain 36.37 per cent., or about $\frac{12}{32}$ (by weight) of salt. When the water drawn from the boiler is found on trial to approach that degree of saltness, it is considered unsafe longer to defer the operation of blowing. By observation and experiment, engineers learn to time it, so as to avoid waste of fuel on the one hand, or a deposit on the other.*

44. If, by not blowing off saturated water often enough a deposit forms on the flues in the shape of a hard incrustation, being a non-conductor, it produces an increased consumption of fuel without an increase of steam, and occasions the flues to burn.

45. The evils finally resulting from inattention to the blow cock in marine boilers are then—first, burning and weakening the flues, especially if of copper, and consequent danger of explosion;—second, an eventual waste of fuel, which, on long voyages, or in cruising vessels, is of great importance. As blowing off renders a copious supply of feed water necessary, and as this feed water is low in temperature and tends consequently to reduce the temperature of the water in the boiler with which it mingles, there may be a necessity for blowing and feeding so freely as temporarily to check the formation of steam. In such cases, steam must be sacrificed to preserve the boilers. True economy in the long run, lies in blowing liberally. But false economy often prevails, and as a consequence, hands are often seen in popular marine steamers with

---

* The hydrometer which marine engineers use, termed a salometer, has its stem graduated to 32ds. In fresh water the instrument will sink to 0, or zero; in sea water, $\frac{1}{32}$ part of which by weight is salt, it will sink only to $\frac{1}{32}$; and will sink less and less deep, as by evaporation the water is made more and more salt; until the point of greatest saturation, $\frac{12}{32}$, is reached, when the water can become no denser, but solid salt will deposit. It is usual to blow when, (if the trial water is at 200° Fahrenheit,) the instrument swims at $\frac{2}{32}$: for although when it swims at $\frac{3}{32}$ salt does not deposit, lime will, if present. Is not "scale" re-soluble?

So the salter the water, the higher the temperature at which it boils. Thus under an atmosphere, at 0 (fresh), it boils at 212°; when of $\frac{1}{32}$ in saltness, it boils at 213.2°; of $\frac{2}{32}$, at 214.4°; of $\frac{3}{32}$, at 215.5°; of $\frac{4}{32}$, at 216.7°; and of $\frac{12}{32}$, at 226°.

The reader is referred to the instrument, as explaining itself better than a picture, engraved or lithographed.

chisels and bars, chipping the "scale" from the flues of the boilers.

46. It is not unusually a fair subject of comment, that engineers often confine their attention wholly to the starting bar, injections, throttles, and other parts of the engine, and leave the water gauges, feed, and blow cocks of the boiler, entirely to the judgment and management of firemen; when in point of fact the boilers, being the principal source of danger, should receive the frequent attention of the best intelligence, and greatest responsibility, amongst those on watch in the steam department of a vessel.

VOLUME OF STEAM—COMPUTING THE EVAPORATION.

47. The "volume of steam," as a specific term, signifies the relation which a given bulk of steam bears to the bulk of water evaporated in generating the steam. In other words, it is the volume of steam produced, as compared with the volume of water which produces it, or to which it will return when, by abstraction of the heat, it is completely condensed.

48. The "volume" of steam decreases with the increase of its pressure, as will appear by inspection of the table of volumes, in which steam of 212° and one atmosphere, is shown to have a volume of 1700; signifying that one cubic inch of water will produce 1700 cubic inches of steam at that pressure; or that $\frac{1}{1700}$ part of steam at that pressure, is water which has been evaporated to produce the steam. At two atmospheres, (one in excess,) temperature 252°, the volume of steam is 883; at three atmospheres, (two in excess, or by the gauge,) the volume is 622; and so on, as per the tables.*

* See Haswell's Engineers' and Mechanics' Pocket-Book, page 208, Ed. 1860.

49. Therefore, to compute evaporation, find out the cubic quantity and the average pressure of the steam (in excess of the atmosphere) used in a given time, and divide the quantity by the volume due to that average pressure. The quotient will be the water evaporated; also the water of the condensed steam.*

---

* To illustrate the utility of knowing volumes, by which evaporative efficiency of the boilers with a given consumption of fuel may be determined, take the Wyoming's performance, reported as follows in the Journal of the Franklin Institute, Vol. 68, No. 407, p. 351:

Mean pressure of the steam on entering the cylinder was (above the atmosphere) 22.8 lbs., the "volume" of which is 717. Point of cutting off 15.6 inches from the cylinder head. Number of cylinders, two; and consequently number of charges of steam used in each revolution 4. Area of cylinder 1964 square inches. Number of revolutions per minute 80.5. Therefore, multiply 15.6, 4, 1964 and 80.5 together, and the product will be the cubic quantity of steam used, in inches. Divide this product by 1728, the cubic inches in a foot, and it gives the cubic feet of steam used in a minute.

Hence $\dfrac{15.6 \times 4 \times 1964 \times 80.5}{1728} = 5709$ cubic feet of steam used per minute.

If this be divided by 717, (the volume,) it will give the *water* evaporated, in cubic feet. And this again multiplied by 62½, the pounds weight of a cubic foot of water, gives the weight of water evaporated per minute.

Hence again, $\dfrac{5709}{717} \times 62.5 = 462.75$ pounds of water evaporated per minute.

The coals consumed equalled 49.5 pounds per minute.

Therefore $\dfrac{462.75}{49.5} = 9.32$ pounds of water per pound of coal, as the rate of evaporation by the Wyoming's boilers, which are of the Martin's vertical water tube description, (Art. 7, Sec. 3, Fig. 3). The fuel used in this experiment was the "Broad Top," a semi-bituminous, free burning coal, mined in Pennsylvania.

Another trial in the same vessel, with 15.3 lbs. pressure of steam (volume 880), cutting off at 12.1 inches from head of cylinder, and making 72 revolutions per minute, gives $\dfrac{12.1 \times 4 \times 1964 \times 72}{1728} = 3960$ cubic feet of steam used, or $\dfrac{3960}{880} \times 62.5 = 281.5$ pounds of water evaporated per minute.

The coal burned per minute averaged 29.15 pounds. Hence $\dfrac{281.5}{29.15} = 9.65$, the pounds of water evaporated by a pound of coal.

SECTION II.

Theory of Latent and Sensible Caloric as applied to Steam. Specific heat.

1. There is, in the formation and condensation of steam, another class of effects produced, which, though less strikingly apparent, are of great consequence because of their immediate practical connection with the subject, as well as their great general philosophic interest: those effects which are explained by the theory of latent and sensible caloric. This theory was first taught by Dr. Black at Edinburgh, in the year previous to that in which Mr. Watt commenced his labors upon the steam-engine, in the latter half of the last century, and served him in explaining the difficulties which he encountered at the very threshold of his work.

2. Heat, in the ordinary acceptation of the term, expresses something which is *sensible*—that is perceived by the sense of feeling, or by that of sight in its effects upon the thermometer. But Dr. Black discovered, that the principle of heat existed in a state not sensible, as well as sensible. To avoid ambiguity, this principle is denominated *caloric*.

Caloric is, therefore, either free and sensible, or latent and insensible. Caloric is known to be the cause of fluidity; and the absence of caloric the cause of solidity. Apply heat to ice or iron, and they become fluid;—cool them, and they resume their solid form.

3. The theory is, that all solid bodies are composed of particles of matter held together by the attraction of cohesion; but that a portion of caloric is interposed

32   LATENT AND SENSIBLE CALORIC.

between these particles, so that they do not ever actually touch—though appearing to; that when more caloric is applied, the particles are separated by it, or to receive it, which causes an expansion, as all bodies do expand by heat; that when, by a further application of caloric, the particles, to receive it, have separated so far asunder as sufficiently to weaken the attraction of cohesion, the body ceases to be solid, and passes into the fluid state, in which the particles move freely amongst one another.

4. Water is taken as an exemplification of Dr. Black's theory of latent and sensible caloric, because most convenient, and because appertaining more immediately to the subject treated.

The substance of which water is composed, exists under three different manifestations—the solid—the liquid—and the æriform. As a solid, it is ice;—as a liquid, water;—and as æriform, a vapor.

5. If ice, with a temperature of about 32 degrees, has applied to it 140 degrees of additional heat, the ice becomes water, with a temperature of a little over 32 degrees. The 140 degrees of sensible heat applied in this case to produce fluidity has all disappeared—become insensible. The heat is in the water, but it is there in a latent state.

To show this—take a pound of water at a temperature of 172°, and put into it a pound of ice at the temperature 32°. The ice will melt, and the result will be, two pounds of water at a little above 32 degrees. The ice has, therefore, taken up 140 degrees of sensible heat in becoming liquid, and the whole of it has passed into the insensible or latent state.

6. Water becomes steam, under a pressure of one atmosphere, at a temperature of 212°; but in becoming steam it takes up an additional 990° of heat;

so that it contains 1202°, of which only 212° remain sensible.

To show this—connect two tightly covered vessels, by a steam pipe passing from the top of one to near the bottom of the other. In the vessel A, put 5½ pounds of water at about 32°; and in B, put an in-

definite quantity. Boil B, and its steam, at 212°, passing into A, and condensing there, will impart heat enough to boil the water which A contains. When the water in A has by this process been made to boil, it will be found to have increased in weight to 6½ pounds; showing, that one pound of water has passed in vapor from B to A, and when in that form held caloric enough to raise 5½ times its weight of water, 180 degrees in temperature—(from 32 to 212)—still retaining its original temperature of 212°.

Hence, as the one pound of water which has passed from B to A in the form of steam, has imparted 180° of caloric to each one of the 5½ pounds of water in A—that is, imparted $5½ \times 180° = 990°$,—and as that one pound still retains all the sensible heat it at any time had, viz. 212°, which it retains as part of the 6½ lbs. boiling in A; it follows, that the said one pound must have taken up and held 990° of latent insensible heat not directly measurable, and 212° of sensible heat directly cognizable by the feelings and measurable by the thermometer;—or a total, latent and sensible, of $990 + 212 = 1202°$.

This experiment shows also, that 5½ pounds of ice-cold water condenses the steam produced by the evaporation of one pound; and will serve in explaining hereafter, why so much water of a higher temperature is necessary in the condenser of the steam-engine, to condense the steam, and produce a vacuum.

7. In general, all reduction of bulk, or increase of density, is accompanied by a development of sensible heat. A direct and vivid example of this occurs when atmospheric air is suddenly and powerfully compressed, exhibiting luminous heat, or producing combustion.

8. Thus also, steam of very great elasticity or pressure, being in effect the same thing as low steam compressed into a smaller compass, as in all other cases of reduction in bulk is accompanied by a great development of sensible, and corresponding reduction of latent heat. And in all temperatures of steam, when not superheated, the sum of the latent and sensible heat is the same (constantly 1202°), high steam developing the latter in greater, and low steam in less relative proportions.

9. On the other hand, an enlargement of bulk or decrease of density is ordinarily accompanied by an absorption of heat from the sensible to the latent state. A striking example of this is found in the fact, that steam at 212°, emitted from a boiler, will scald; when steam of 350°, under like circumstances, will not scald. In one case, there is so little expansion in bulk when released from the confinement of the boiler, as to absorb and render latent but little of the 212 degrees; in the other case, the expansion is immense and rapid, absorbing and rendering latent more than 150 degrees of the sensible heat, reducing it below the scalding temperature.

10. Thus again, when, as will be explained further

on, high steam is expanded in the cylinder of an engine, the expansion is accompanied by great absorption of sensible heat into the latent state. But the sum of the latent and sensible heat remains the same (a constant quantity) after the expansion as before the expansion—the difference being only, that after the expansion less of the heat is sensible, and therefore to the same extent more of it is latent.

11. And, inasmuch as the condensation of steam is produced by extracting its heat, usually by means of condensing water (Art. 7); and as the proportion of condensing water required depends upon the heat to be extracted; and as that heat amounts to the same whether more or less of it be latent, in other words, whether the steam is expanded or not; it follows, that in theory at least, the condensing water required is in proportion to the quantity and force of the steam admitted to the cylinder, whether that steam be there expanded or not expanded. Practically, in so far as expansion of the steam reduces the temperature or sensible heat, and causes condensation, less condensing water may be required by the expanded steam.

SPECIFIC HEAT.

12. Dr. Black's investigations in heat, led to another discovery—viz., that different substances have different capacities for receiving heat yet undergoing only an equal change of temperature;—that is to say, to raise equal weights of two different substances 10° each in temperature, one may require twenty times more heat and fuel than the other. For example, to heat water, oil, and mercury, will require different quantities of heat, in the proportions of 23 for water, 11.5 for oil, and 1 for mercury; or, assuming, as is

usual in the tables, the heat required to raise the temperature of fresh water to be the unit, water will stand as 1.000, oil as $\frac{1}{2}$ or .5, and mercury $\frac{1}{23}$ or .0435 (decimal). This difference of capacity for receiving heat, found to exist in various substances, is designated as their "specific heat;" and in the tables of specific heat, that of various substances is registered as determined by experiment; and that of water will be found so registered as 1.000—of oil as .5—and of mercury as .0435.

The theory of specific heat is principally of use in relation to steam, because super-salted water has a different capacity for heat from fresh water; and requires, in proportion as it is super-salted, a different amount of heat to raise it a given number of degrees in temperature.

SECTION III.

Construction of Boilers. Fire surface. Water room and Steam room. Forms of boilers, and kinds of metal used in their construction.

CONSTRUCTION OF BOILERS.—FIRE SURFACE AND FLUES.

1. The efficient action of a steam-engine depends upon a supply of steam of such elastic force, and in such quantities, as the engine was designed to consume. (Sec. 1, Art. 4.) If the estimated effect of an engine be based upon the hypothesis of 20 revolutions, or a supply of 40 cylinders of steam per minute, having a pressure equal to 10 pounds per gauge, and the boiler prove adequate to supply steam of that average

pressure in the cylinder at a rate to produce only 10 revolutions per minute, it follows that the engine will act with much less effect than was designed, and will imperfectly execute the work intended.

2. The correct proportions and construction of a boiler become, therefore, of the highest consequence; and its first requisite is to supply steam to the engine at its calculated number of revolutions.

To accomplish this, the furnaces, besides *capacity*, must possess good *draft*, so as to produce perfect, vigorous combustion and give intensity to the heat. And the boiler must be constructed with a sufficient extent of *fire surface* (Sec. 1, Art. 13), for boilers may make steam of a force and with a rapidity, proportioned to the capacity of furnace, the draft, and the extent of fire surface.

3. The intensity of the *draft* depends upon the height of the chimney, or pipe; and, in a minor degree, upon the *area* and *arrangement* of the flues.

4. By the *area* of flue is meant the area of a transverse section, or, when there are several flues, the sum of the areas of all the sections. This area is generally left to result as it may, from the arrangement for producing a sufficient extent of fire surface; presuming when that is obtained, the area of flue will be also sufficient.

5. The *arrangement of flue* is made so as to take up in the water *all* the heat produced which is of a temperature high enough to make steam. That is the principle. The practice cannot always conform. But the endeavor is to approximate as nearly as possible; for manifestly, to permit a large proportion of this heat to pass off into the air through the chimney, would be a waste of heat, *which is a waste of fuel.* Nevertheless, in spite of all possible perfection in the arrange-

ment of marine boilers, one-third of the heating power which exists in coal is supposed to be lost by incomplete combustion and the escape of heat up chimney. (See Isherwood's incomparable work, Vol. 2, p. 49.)

This escape is reduced, by extending the fire surface, either by passing the flame and heated air through a large number of small short flues (Art. 7, Fig. 2); or through a small number of large flues of greater length (Art. 7, Fig. 1); or among vertical water tubes, as in Martin's boiler (Art. 7, Fig. 3).

6. It is thought better, especially for marine purposes, if flues are used, to have them not less than from 8 to 12 inches diameter; provided there is room for boilers long enough to give the length which large flues must have in order to produce the requisite fire surface; or high enough to produce this length by returning the flues.

The large flue is preferable, because less liable to choke with soot, ashes, cinders, or salt which comes from leakage. But in situations which restrict length, height, and width of boiler, the only method of producing in a flue boiler such extent of fire surface as will extract all the heat capable of being used to advantage in generating steam, is to reduce the size and multiply the number of flues, and take the attendant disadvantage of liability to choke. Boilers in which the diameter of flues is reduced to 3 or 4 inches, are termed tubular boilers (Art. 7, Fig. 2); and on account of their reduced size and weight, are universally in use for producing locomotive power on railroads. The Figures 1, 2, and 3, following, illustrate nearly all that has been or will be said in this Section respecting the form and construction of boilers.

CONSTRUCTION OF BOILERS. 39

7. Fig. 1 is a longitudinal, and a transverse section of an ordinary low pressure American angular flue

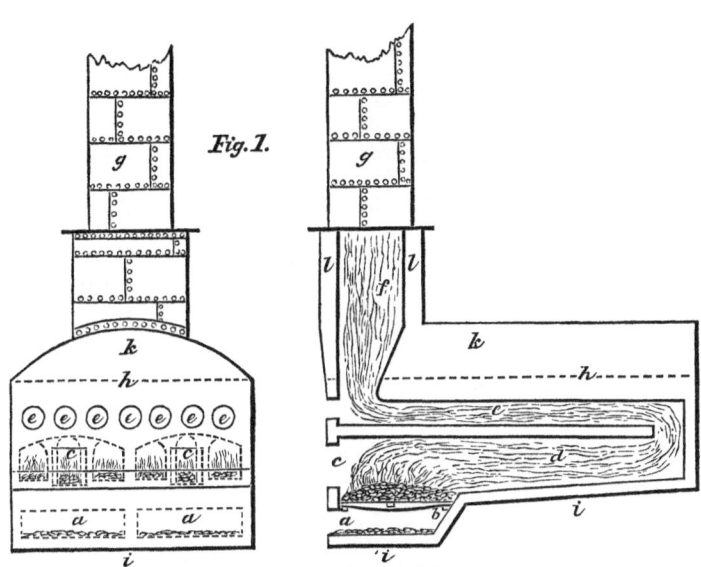

boiler; in which a is the ash pit, b the grate bars, c the furnaces, d the flues, e the return flue,* f the flues united, g the smoke pipe, h the water level, i the water bottom, which protects the bottom of the boiler from burning, and renders available some heat that would otherwise radiate and be lost, k the steam room, and l the steam chimney in which steam is slightly surcharged or superheated, and from which the steam is taken off to the cylinder of the engine.

Fig. 2 is a longitudinal section and front view of a tubular cylindrical high pressure boiler; in which a is the ash pit, b the position of the grate bars, c the fur-

---

* Evidently if in Fig. 1 the flue did not return, but passed into a chimney through the back end of the boiler, a vast amount of heat would be lost. In some boilers there is a *second return* of the fire through the length of the boiler, before it passes into the chimney, in which case the chimney is over the back end.

nace, *d* the tubes corresponding with the flues in Fig. 1, *e* the smoke box or base of the chimney in which

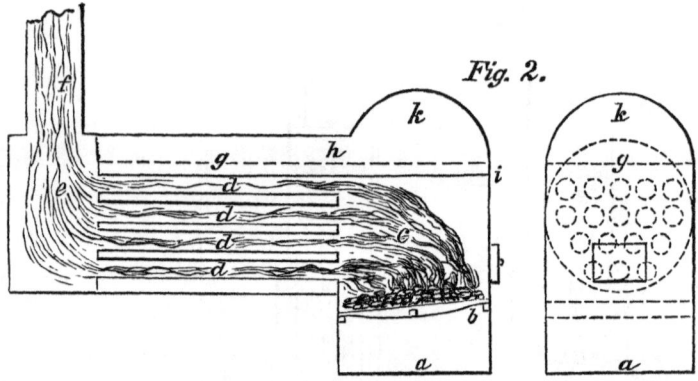

the tubes terminate, *f* the smoke chimney, *g* the water level, *h* the steam room, *i* the top line of the furnace and flues, and *k* the *dome* from which steam is taken off.

MARTIN'S VERTICAL WATER TUBE BOILER.

This description of boiler is said to possess advantages greater than any other for marine purposes. (See Isherwood, Vol. 11, pp. 164–183.) The figure 3, following, represents its general features, *a* being the ash pit, *b* the grate bars, *c* the furnace door, *d* the flue, *m* the back connexion, and *e e* the return flue, in which the vertical water tubes are set. They contain water circulating vertically through them between *t* and *h*, *t* representing the water over the crown of the furnace, and *h* the water surface in the boiler. The draft from the furnace, and through the return flue *e e*, passes amongst these vertical water tubes, and imparts heat to the water in them and circulating through them.

Though the disposition of this boiler, like any other, may be lengthwise with, and the fire room across the ship, yet economy of room is thought to be promoted by the present usual disposition, which lays

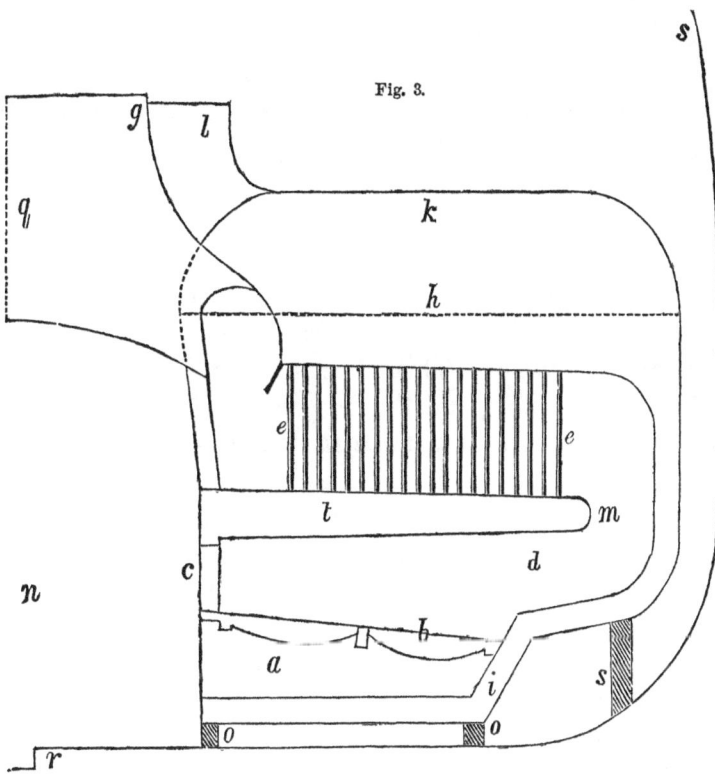

Fig. 3.

the flues, grates and furnaces crosswise with the ship, and opens all the furnace doors of both sets of boilers, one set on each side, into a common fire room *n*, which being lengthwise with and over the keel *r*, is, though warmer by radiation, cooler by the more airy circulation; is more handily coaled; is attended by a single gang of firemen; is more under the eye of the engineer in charge, and brings all the flues more readily into one smoke stack *g*. Around the smoke stack *g*, is the steam chimney *l*, which is a half drum common for all the steam made by the furnaces of one side of the ship, which half drum is divided from the half drum of the other side, by the partition q. The water bottom, as in Fig. 1, is i; the ship's bottom $o\,o$; the side s; and as in the others, k is the steam room.

The rolling of small vessels is said to throw the coals out of the furnace doors, and the disposition in question for that reason to be objectionable.

WATER ROOM AND STEAM ROOM.

8. The next consideration of importance in the construction of boilers, is their capacity to contain water and steam. This of course depends upon the size of the boiler, and the proportion of space within occupied by flues—for if the space in a boiler be nearly filled with flues, as is the case with tubular boilers, there can be but little room left for water.

The boiler represented by Fig. 2, being tubular, has, as compared with one like Fig. 1, most fire surface in the same sized shell of boiler—but the water room is very much reduced by crowding the flues so close together.

9. On shore, there is rarely any limit, other than cost, to the increased size or weight of a boiler, and it is easy, therefore, to construct one for a land engine, with all the flue necessary, and with abundant capacity for both water and steam. But in designing a boiler to be put afloat, the engineer is restricted both in room and in weight; for if a vessel is occupied and laden with boilers and engines, she has neither room nor ability to carry any thing else.

10. The reduction of water to be carried in a boiler afloat, is therefore highly desirable, and is limited only by the following considerations, which are imperative:

First.—The quantity of water carried must be so much greater than the evaporation in a given time, that the re-supply of cooler feed water in the same time shall not greatly reduce the temperature, and consequently not suddenly check the formation of steam. The water carried must also be so great in proportion to the saturated water formed and blown off, that the consequent re-supply of cold feed water shall not suddenly and greatly reduce the temperature in the boiler.

Second.—There must be such height of water carried above the upper flues, that small neglect of the feed will not expose them bare of water; and the height of water over the flues also such, that when the vessel heels, the outermost flues will not be left bare by the water preserving its horizontal surface.

11. The height of boiler necessary to give *steam room*, (k Art. 7,) as the space in the shell of the boiler above the water is termed, is also reduced as much as the considerations involved will permit.

These considerations are:—first, the necessity of taking off steam at a point so far above the water surface, that no water can work with the steam into the cylinder;* for water being incompressible, any quantity of it between the piston and the cylinder head would burst the head out;—and second, the necessity of giving to steam an accumulation in quantity equal to at least 10 times the capacity of the cylinder; for if the accumulation be less, steam is worked off in so large proportions as to relieve pressure from the surface suddenly, and produce violent ebullition or foam-

* The tendency of water to work into the cylinder with steam, is very great, even where steam is taken off much above the water level. It is caused by the sudden relief of pressure at one spot on the surface whilst the pressure continues on the remainder of the surface, acting like suction.

ing, (Art. 33, Sec. 1,) which occasions dangerous deception as regards the height of water in the boiler.

12. When boilers are constructed with a height so limited that steam cannot be taken off high enough above the water level to avoid the danger of working water into the cylinder, there is elevated on the boiler a drum, or dome, (as k Fig. 2) from which to take the steam; or else a steam chimney, (as l Fig. 1) which is a drum constructed about the smoke pipe, extending a few feet above the shell of the boiler. The steam in this chimney becomes, to some extent, surcharged by the heat it derives from the smoke pipe. As before remarked, a principal object of giving to the steam this extra heat, is, that if it lose any by radiation in its passage along the steam pipe, it may not condense.

STRENGTH OF BOILERS, AND KINDS OF METAL USED IN THEIR CONSTRUCTION.

13. Boilers must of course have strength in proportion to the elastic force of the steam they are intended to contain. With a single atmosphere of steam within, balanced by the pressure of the natural atmosphere without, no extraordinary strength of construction is necessary. But when steam forms, as is not unfrequently the case, of 6 and 8 atmospheres pressure, or more than 100 pounds to the square inch, the greatest strength, both of form and construction, attainable under the circumstances, is required.

14. Low pressure boilers, which rarely carry steam exceeding 25 lbs., are usually made in an angular shape, which is favorable to economy of room in ships, and when well braced by rods within, are found strong enough.*

---

* Nearly all American boilers of the angular form, are braced by rods from the bottom to the top, from side to side, and from flue to flue, and from the flues to the shell. These braces are often carried to points not more than six inches apart.

15. But the high pressure boiler, carrying 100 pounds and upwards, is invariably of the cylindrical form, which exceeds all others in strength, and needs no bracing. And as their strength is inversely as their diameters, boilers for very high steam do not exceed 5 or 6 feet in diameter. In such cases, great boiler capacity is obtained by increasing the number of boilers, and uniting them with water and steam connections, so as to preserve an equilibrium of both water and steam throughout the series.

The danger of this latter arrangement consists in the possible obstruction of these connections; and in the liability, when afloat, of an outside boiler having its flue exposed by a small heel of the vessel. This often happens on western boats that carry little water in their boilers, and in the bend of the rivers take great heel.

16. Generally it may be known from the position of the chimney, over the front or over the back end, if a marine boiler is a flue boiler, with or without tubes set in the return flue (see Figs. 3 and 1, Art. 7), or a tubular boiler (Fig. 2, Art. 7); which will appear by inspection of the Figures.

Small "donkey" boilers, usually for economy of space have flues standing vertically, which position has very many advantages in all cases, as is lucidly pointed out by Mr. Isherwood.

17. Boilers are constructed with plates of metal, rolled out from copper or iron, and riveted together. Copper boilers as compared with iron, are more costly, not so strong, and more liable to injure or burn out by exposure to fire when not protected by water. On the other hand, they corrode less, and for that reason, especially if used in salt water and carefully attended, may last much the longest time. And when worn

out, the copper boiler is highly valuable as old metal, whereas the old iron boiler is nearly useless.

It may be observed, that iron boilers are coming into use, to the general if not entire exclusion of copper.

SECTION IV.

Combustion. Draft. Blast. Fuel. Smoke.

COMBUSTION.

1. Combustion is a rapid oxidation and decomposition of substances, attended by great development of heat, which is generally, though not necessarily luminous. This decomposition results from the affinity which the combustible substance has for oxygen. The stronger this affinity, the more combustible the substance. Sometimes the affinity is so great, as to produce spontaneous combustion; but usually, combustion does not take place unless excited by the application of high artificial heat.

2. Where the supply of oxygen is most rapid, the combustion, and consequent development of heat, are also most rapid. And as the natural supply of oxygen is derived directly from atmospheric air, 20 per cent. of which is oxygen, it follows, that the more rapid the supply of air to a furnace where combustion is going on, the more vigorous will that combustion be.

DRAFT.

3. This supply of air depends upon the draft; which, other things being equal, is proportioned, within extended limits, to the height of the chimney or pipe.* If the column of air in a chimney is not heated, its weight will be equal to that of a column of air of equal area, whose base is without the chimney; the two columns, therefore, balance each other, and no motion or current is perceived. But if that portion of the atmospheric column whose base is in the chimney, be heated, it is rarefied, and becomes lighter than the surrounding columns; and as the air always moves in the direction of least resistance, a current more or less strong immediately arises in the chimney. The higher the chimney, the greater the rarefied portion of the atmospheric column resting in it, consequently the greater the current, and more powerful the draft. And if this current passes through the furnace grates, it yields its oxygen to the fire there, and produces a vigor of combustion proportioned to the draft.†

BLAST.

4. Some kinds of fuel in common use, as anthracite coals, burn better with more air than can conveniently be obtained by the height of chimney in steam vessels. In such cases, an artificial blast is sometimes resorted to, and this blast is generally produced by means of a fan blower, which makes several hundred

---

* A chimney may be so high that friction will obstruct draft.

† The intensity of the draft is as the square root of the height of the chimney. In locomotives, great intensity of draft with a short smoke chimney is produced by exhausting steam into the chimney. This simple expedient has rendered the locomotive complete for its purposes.

revolutions per minute, and forces atmospheric air under the grate bars of a furnace, and up through the fuel spread upon them.

FUEL.

5. Bituminous, semi-bituminous and anthracite coal, pine and hard wood, are the kinds of fuel in common use.

6. The components which in fuel are essential to combustion, are hydrogen and carbon. Hydrogen is the principle which produces blaze; carbon that which constitutes the hot coke or coal bed, that radiates heat after the inflammable principle is exhausted. That fuel, therefore, which contains the largest proportion of hydrogen, as bituminous coal, pine wood, tar, and rosin, will make the most diffusible, but least enduring fire; and that which contains the largest proportion of carbon, as anthracite and semi-bituminous coals, produces the most enduring, though a less diffusible fire. If it be desired, then, to make a sudden and intensely diffusible heat, it is common to throw resin into the furnaces; but evidently, this will not make a lasting fire. On the other hand, anthracite coal fire is very enduring, but under a natural draft, although possessing high local heat, does not so well diffuse it through an extended set of flues. The fuel which best combines the inflammable and enduring principles, is the semi-bituminous coal, as the Cumberland Maryland coal; and is, therefore, usually preferred.*

7. In selecting wood for fuel, get that which is sound and dry. In selecting coal, get that which is most free from slate and other earthy matter, which

---

* See Haswell—title Combustion—page 224.

always combines in a greater or less degree in coal, and forms the incombustible residuum termed ashes, and which when fused under great heat, becomes clinker. As this earthy matter is paid for as coal; weighs, and occupies space on board ship; chokes up the grate bars, impairing draft; absorbs heat uselessly; causes the furnace doors to be opened for cleaning the grates, which admits cold air to the flues; and makes labor in cleaning out and hoisting the cinders overboard; of course, the less a body of coal has of it the better.

Select also coal which does not crumble readily, because fine coal wastes with the ashes through the grates; and it cakes on the grates so as to impair draft unless broken up often, which causes the doors to be opened, admitting cold air; and when it is broken, passages are often made too large, through which air passes into the flues not deoxygenated, and inadequately heated. All bituminous coals have this disadvantage to a greater or less extent. But usually they make perhaps less ashes than the anthracites, and have the advantage of being free burning, by which the fires may be kept more even.

Stowage also is an important consideration, for some coals occupy more, and others less room, in proportion to their heating properties. The reader is referred, for a full discussion of the whole subject, to Isherwood, Vol. II., pp. 29–35.*

8. Coal from very many mines contains sulphur. The effect of sulphur, owing to its strong affinity for iron and copper, is to destroy the grate bars, and injure seriously the furnaces and flues where coal con-

---

* For all sorts of tabulated facts respecting fuel and combustion see Haswell's Pocket Book, pp. 224–228., Ed. 1860—a book which no practical person in any position of life should be without.

taining it is burned. The coal, therefore, most free from sulphur, is best for use in boilers.

9. Another important reason why coal containing sulphur should be avoided for steamers is, that sulphur is the principle which, if combined with iron forming the sulphuret of iron, or iron pyrites, occasions spontaneous combustion by the great development of heat which takes place when, in a state of long continued moisture, a change in the chemical union of the sulphur and iron occurs, from a sulphuret to a sulphate of iron. This is an important fact, for spontaneous combustion has often broken out in steamers, and forms a prominent source of danger in them, unless intelligibly guarded against, as is most effectually done by keeping the coals dry. The pyrites, being a light gray mineral, is easily detected by examination. It occurs in some of the Pennsylvania and Virginia bituminous coal mines; but in the Maryland coal region has not yet been detected.

SMOKE.

10. The product of a *perfect* consumption of the combustible ingredients in fuel (Art. 6), is carbonic acid, and watery vapor; the oxygen combined with the carbon producing carbonic acid, and with the hydrogen producing vapor. These products, both of which are incombustible and invisible, are carried off by the draft into the chimney. But *imperfect* combustion sets free from all kinds of fuel in common use, except anthracite coal, large quantities of combustible matter, chiefly carbon, which escapes as *smoke* through the chimney into the air, and is lost. Frequent attempts to burn the smoke have been made, but never until recently with success. A former method which

promised well, was to admit a fresh current of air to the flues, behind the bridge wall. The oxygen of this fresh supply of air combining with the carbon, and whatever else of a combustible nature exists in the smoke, consumes it, developing at the same time a new supply of heat. But less heat was found to be gained by this new combustion, than was lost by admitting to the flues the cool air which went to produce it.* Considerable success has attended a recent practice of perforating the furnace doors, so as to pass a moderate current of air over the coals, which is found to promote combustion and the production of steam, and in some degree reduce the smoke.†

11. But even if nothing in economy of fuel is gained by consuming the smoke of bituminous coal, to be rid of it is certainly a great convenience. And to cruising steamers of war, it might be of essential importance not to have their positions exposed by smoke floating over them.

* These principles of draft, and perfect and imperfect combustion, and the consequent absence or presence of smoke, are illustrated by the Argand burner commonly used on parlor tables, and known as the Astral lamp.

Light this lamp without the chimney, and in consequence of imperfect draft the combustion will be imperfect, and the lamp will burn with a dull red flame, yielding quantities of smoke. Put on the chimney without altering the wick, and the improved draft will occasion the flame to assume a bright white appearance, without smoke,—showing a perfect combustion. Turn the wick higher, and the red flame and smoke will again appear, because the fuel is increased disproportionately to the draft or supply of oxygen, and the combustion is again in consequence imperfect, (Art. 3.)

† The Argand principle, is simply to give the burning wick of a lamp a double supply of air, one supply within and up through the tube on which it is fixed, and the other supply from without. Stop the holes below, or those at the side, through which the supplies are received, and their utility will be rendered apparent by the red flame and smoke which ensue. And the admission of air over the furnace grates of a boiler, acting with the supply from beneath them, which places the fuel between two separate supplies of air, is but another, yet very ingenious application of the principle illustrated by the Argand lamp.

SECTION V.

The steam engine. Interior construction of the cylinder and condenser. General description of the engine. High and Low pressure. Surface condensation Heaters for feed water. Expansive action of steam.

THE STEAM ENGINE.

1. The engine is worked by steam which flows from the boiler into the cylinder, where it acts upon the piston and produces motion. The steam is admitted to the cylinder through a valve, and when the steam so admitted has performed its work of driving the piston either up or down, another valve opens and permits this steam to escape from the cylinder. In some instances (as in the high pressure engine), this escape of steam is into the air; in others (as in the low pressure engine), it is into a vessel called a condenser, where the steam mingles with cold water, and is condensed. The escape into the air is simple and direct. But the escape into the condenser, and the operations there, require explanation.

INTERIOR CONSTRUCTION OF THE CYLINDER AND CONDENSING APPARATUS.

2. The interior construction of the cylinder, piston, condenser, and air pump, and their mode of operation, are represented in the adjoining Figure. Beginning at the top of the figure, *a* is *the piston rod*, passing through the *cylinder head b*, by an aperture rendered steam tight by a *stuffing box c*, which screws down and

packs hemp at *d*, tightly around the rod. The *piston e*, is also packed steam tight by hemp or metallic pack-

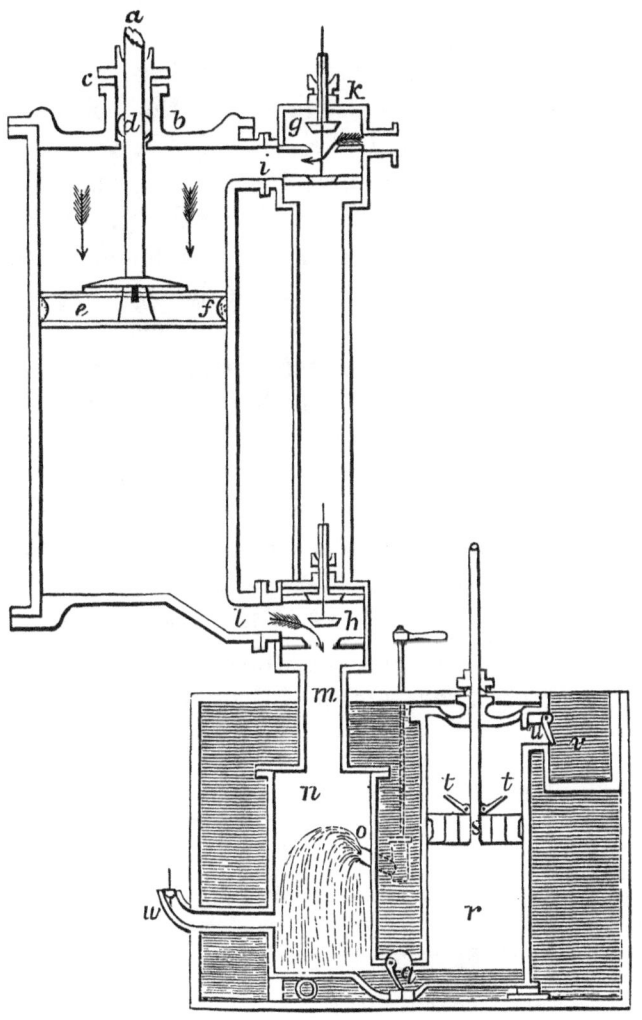

NOTE.—This figure represents the entrance of steam to the upper nozzle, and escape from the lower.

An opposite view of the *side pipes*, *steam chests*, and steam and exhaust valves, would show with equal clearness, the entrance of the steam at the lower nozzle, and its escape from the upper nozzle into the condenser, together with the upward motion of the piston.

ing at *f*, which prevents steam passing between the piston and cylinder. The upper *steam valve g* and the lower *exhaust valve h*, are opened at the same time; and then steam flows in at the upper *nozzle i*, from the *steam chest k*. This steam chest is kept constantly full of steam by a pipe leading from the boiler. As the steam flows through the *nozzle i*, it forces the piston down in the direction of the arrows; and at the same time the steam that is below the piston (after having forced it up) flows out of the cylinder through the *nozzle l*, the *exhaust valve h*, and the *exhaust pipe m*, into the *condenser n*. Here the steam meets and mingles with the injection water flowing in through the *injection pipe o*, imparting its heat to the water, and becoming itself thus condensed to water, which makes a vacuum.* The condensation of this steam produces a nearly perfect vacuum below the piston; consequently there is but little resistance to the downward motion of the piston, and the whole pressure of the steam upon its upper surface is *effective* pressure—pressure per gauge plus the vacuum. When the piston has reached the bottom of the cylinder, the valves *g* and *h* close, and other valves (also seen in the

---

* When a space is filled with steam to the exclusion of every other substance, remove the steam, as is done by condensation, and a vacuum is the immediate and necessary result. A cubic foot of steam under an atmosphere, when condensed to water, occupies the space of a cubic inch, (or nearly that, for strictly 1700 not 1728 is the volume under an atmosphere,) in which case 1727 inches of the space would become vacuum. If the vacuum of an engine were produced by exhausting the air through any mechanical force, as with an air pump, the power obtained by the vacuum, would be less than that expended in obtaining it, by exactly the amount of friction. Like pumping water to the top of a hill for producing an artificial water power—it would cost more than it would come to, by the amount of friction expended. But by the condensation in question under an atmosphere, we have a cubic foot of vacuum, with a base of 144 square inches, by which a force of $144 \times 15 = 2160$ pounds is obtained, and the mechanical cost of it is only that required to pump out, say 50 cubic inches of condensing water, drawn though against an atmospheric weight greater than suction pumps usually experience. See foot note to next page.

figure) open. Steam then flows in through the *nozzle l* to the cylinder, and forces the piston up again, while the steam that is above the cylinder now escapes through the *nozzle i*, and through the *exhaust side pipe* (not visible from this side view) into the condenser. And thus is produced the alternate motion of the piston rod *a*, which connects with a beam or other working part of the engine, as will be shown hereafter.

The condensing water collects in the bottom of the condenser and *bed plate*, and is drawn thence through the *channel* of the bed plate, *and foot valve q*, into the *air pump cylinder r*, by the motion of the *air pump* piston *s*. This piston, though sometimes solid, has commonly valves as *t t*, which open upwards when the piston descends, and shut when it ascends, raising water and forcing it through the *delivery valve u*, into the hot water cistern or *reservoir v*.* The arrangements of the condenser, air pump, and reservoir, are various, but are all upon the same general principle, and with the same object, viz., to produce a vacuum in the condenser, and that part of the steam cylinder which is open to the condenser.

The valves *g* and *h*, in the preceding diagram, are of the kind called *puppet* valves. The puppet valve is a conical plate which lifts from a circular seat.

3. The alternate admission of the steam above and below the piston is also effected by means of the *slide* or D valve. This sort of valve is used on all the high pressure engines in this country; in England for low

---

* It may be observed that the air pump is always level, or nearly so, with the condenser, because otherwise, for the want of atmospheric pressure to aid or rather cause suction, the air pump would not draw, and cannot against any considerable gravity. Indeed, the condensing water rather flows by its gravity through the foot valve *q*, and is *lifted* overboard, or with a solid piston would be *forced* overboard. There is very little *suction* in the case, but yet heavy work for the air pump.

pressure paddle wheel engines; and very generally, if not universally, for screw engines everywhere.

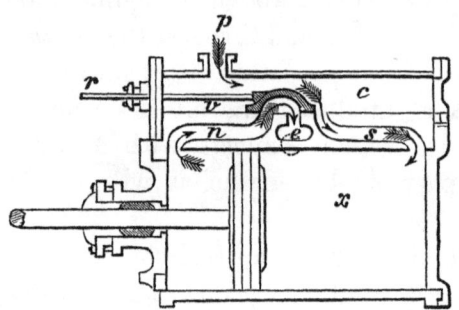

Let x be a cylinder, c the steam chest, p the steam pipe, e the exhaust pipe, and n and s openings which connect the nozzles at the extremities of the cylinder with the steam chest, and v the D valve, which moves back and forth nearly its length in the steam chest, by means of the rod r, worked by the eccentric. In the position represented by the figure, the opening n connects with the exhaust e, and the steam from the upper end of the cylinder escapes; whilst the opening s, being open to the steam chest, receives steam, and conveys it to the other end of the cylinder, as the arrows in both cases show. In the reverse position, s would be connected with the exhaust e, and n be open to receive steam from the chest c.*

GENERAL DESCRIPTION OF THE ENGINE.

4. The reader is now prepared to comprehend the action of the engine as a whole. The plate represents

---

* This D slide valve, bound down to the surface on which it slides back and forth by the steam over it and the exhaust under it, would, in a large valve, have a destructive amount of friction, but for a compensating or balancing device by which it is greatly relieved. Any one with a desire for knowledge on this subject, can readily learn by inquiry what this device is.

the form of a condensing engine, such as is most common in our river passage boats. The parts taken up in detail, and in the order that steam from the boiler reaches them, are as follows: The *steam pipe a*, takes steam from the steam drum at a high point, (Sec. 3, Art. 12,) and conveys it to the *steam side pipe b*,* whence it passes into the *steam chests*, one of which is seen at *c ; d* is the *cylinder ; e* the *exhaust pipe ; f* the *condenser ; g* the *injection cock ; h* the *injection pipe ;* † *i* the *channel ; k* the *foot valve ; l* the *air pump ;* and *m* the *hot* water cistern. The operations in these parts have already been explained in Art. 2. What water is necessary to feed the boiler, and re-supply that which has passed out of it in steam, is taken from the *hot water cistern* through the pipe *n*, by the *force pump o*, and thrown through the *feed pipe p*, and *stop valve q*, into the boiler. This stop valve rises on the downward stroke of the force pump piston, and permits water to pass into the boiler. But on the upward stroke of the force pump, the stop valve falls, and prevents the pressure in the boiler driving water back down the feed pipe. The greater part of the water in the hot water cistern passes off by the waste pipe, through either the side or bottom of the vessel.‡

* In the figure, the steam side pipe is not seen because in this side view of the engine, it is immediately behind the exhaust side pipe, and the steam chests.

† All marine steamers are provided with two—the bottom and side injections. If a boat runs in very shoal muddy water, or grounds, the bottom injection may choke with mud, in which case the side injection is available. Float wood, ice, &c., not unfrequently choke the side injection, in which case the bottom one is available. Marine steamers are provided also with a third injection pipe, called the bilge injection, which opens into the hold of a vessel, to be used for freeing her from water in case of a great leak occurring. The bilge injection together with the bilge pump will free an enormous amount of water.

‡ As will hereafter be more fully shown, the condensing water admitted to the condenser is many times, in some instances from 60 to 70 times, greater than the quantity required to feed the boiler. Hence much the largest proportion of water admitted to the condenser and pumped into the hot water cistern, goes out of the vessel by the waste pipe. Refer back to the last paragraph of Art. 6, Sec. 2. p. 34.

58   THE STEAM ENGINE.

5. This circulation from the boiler through the cylinder and condenser to the hot water cistern, and thence back again to the boiler, is continually going on when an engine is in regular operation. And when a new supply of steam is entering the cylinder above the piston, the steam below, which has done its work, is at the same instant passing off into the condenser. So that a steam valve above and an exhaust valve below, or an exhaust valve above and a steam valve below, open nearly at the same instant of time.

6. The reciprocating rectilinear motion of the piston is converted into a rotary motion, by means of the *piston rod r*, which takes hold of the *working beam* at *s*, and produces a vibratory motion in both ends of it; and, in consequence, the *connecting rod t*, taking hold of the *crank pin u*, turns the crank and water wheel.

The rectilinear motion of the piston rod is preserved by means of a cross head and guide rods; which is a simple arrangement, and is to be seen in all our boats. The English use the parallel motion invented by Mr. Watt.* It answers the purpose in short strokes, but not in the long strokes common in this country.

7. On the shaft, near to the crank, is an *eccentric v*,† which, connected by an *eccentric rod w* with an *eccentric pin* on the *rock shaft y*, occasions the valves to work, and the engine, when once in full operation, to continue so without the aid of any agency other than to supply the fires. The *injection handle* 1,

---

* For a detailed account of the parallel motion the reader is referred to Russell on the Steam Engine, p. 201.

† Any one not understanding what an eccentric is, or the meaning of any other term used, can soonest learn it by inquiry and observation. This book is planned indeed to be suggestive of inquiry, and to start a train of thought and investigation, and give it direction, rather than to pursue it.

serves to open the injection cock g. The connection between the handle and cock is apparent in the figure. The throttle valve 2, in the steam pipe a, serves, by opening or closing, to regulate the supply of steam used.

8. Every condensing steam engine has all these various parts from a to y, performing all the offices described in the one here presented. To this remark there is one exception. Some engines have no beams, but the connecting rod t connects directly from the piston rod r with the crank pin u. The forms of engines are so various, that unless they are studied systematically, they will be likely to confuse a mind not familiar with the subject. Therefore, when a person meets any strange form of engine, let him commence its investigation at a, which is the large pipe always leading from the boiler, and follow the alphabet through, and he cannot fail to comprehend the essential peculiarities.

9. The form of engine represented in the figure, is that of the common "beam engine" in use on board most river boats in the United States which condense their steam. But this lofty arrangement though sometimes used does not answer well at sea, because it holds wind, makes vessels crank or unstable, and above all renders the decks so open, that a vessel shipping a succession of seas, might be more apt to founder.

The beam, therefore, of marine engines, if they have any, is placed below, near the bottom of the vessel, and connects with the crank upwards, instead of downwards. Almost all British marine paddle wheel engines carry beams thus below, and are denominated "side lever" engines.

10. There are many forms of condensing marine engines, some of which, with their peculiarities and alleged advantages, will be briefly enumerated;—the

object of the notice being mainly to stimulate attention, and direct observation.

11. First is the "Beam" and "Side Lever" Engines, before mentioned.

The "Square Engine," is that formerly, but not latterly, very much used on rivers, having the cylinder to stand over the paddle wheel shaft, with two cranks, one on each side of the cylinder, and two connecting rods from the cross head to the cranks, working one on each side of the cylinder.

12. The "Direct Acting Marine Engine," has no beam, but the connecting rod, with one end strapped to the cross head, takes hold with the other end on the crank pin. When this direct action is from a vertical cylinder, as in case the cylinder stands *under* a paddle wheel shaft; or when it is from a horizontal cylinder, as in case the cylinder lays on its side athwart ships, and works the crank of a screw shaft;* the stroke is, from want of room, in both cases very short, and the connecting rod also short. But when it is necessary to give a direct action marine engine below decks greater length of stroke, that object is effected by inclining the cylinder; its engine is then denominated an "Inclined Engine."†

---

* Persons are presumed to know that a screw propeller in the stern of a vessel, has a shaft extending from the screw, along over the keel to the engine.

† When there are two engines to the same shaft in a ship, attached to cranks which stand with right angles to each other, one engine is at the half stroke when the other is on the centre or at the dead point; consequently, however slow the engines may be working against a heavy head sea, they cannot stop on the centre, as a single engine under like circumstances probably might—and more probably would if a short stroked engine, than if one with a long stroke. It is this circumstance, together with the fact that a long stroked engine is of the two much more easily handled, which gives it a special importance as a single engine at sea, and in ferry boats, which work much by hand, must answer the bell quickly, and must not be very liable to hang on the centre or dead point.

Less friction, comparatively, on the crank pin, and less liability to heat or cut, is a merit usually attributed to the long stroke—that is, to a given cubic capacity of cylinder, disposed with, relatively, more length and less diameter. '

13. The "Back Action Engine," is one in which, for economy of room, and to lengthen the connecting rod, the cross head and the cylinder are on opposite sides of the shaft. This usually involves the necessity of a double piston rod to a single cross head, from which a single connecting rod works *back* on the crank.

14. The "Steeple Engine" is, however, a form of back action engine with only one piston rod, strapped to the middle of the shortest side of a triangular or harp-shaped iron frame, within which the crank revolves, and from the apex of which a pair of connecting rods work *back* on the crank.

15. The "Trunk Engine," has no piston rod, but is a direct action engine, in which the connecting rod takes hold immediately on the piston itself, through a hollow open cylinder within the steam cylinder. This engine is favorable on the score of room, but increases the size and weight of the cylinder in order to obtain a given surface of piston on which the steam acts, and is said also to increase friction. In the English navy it is apparently a favorite, but not so in our own.

16. The "Oscillating Engine," is one which has no connecting rod, but, with the piston rod, takes direct hold of the crank pin; consequently has a vibrating or oscillating motion on trunnions, set near the centre of the cylinder, through which trunnions, being hollow, and the only fixed non-moving points, the steam and exhaust pass, from the steam chest, and to the condenser. Of all forms, this is, for side wheels, most favorable for economy of room and on the score of friction; for which reason it is a favorite, though it is less frequently met with than some other forms.

17. A "Geared Engine," is one in which, to ob-

tain an increased number of revolutions without an increase in the speed of the piston, cog wheels are introduced, by means of which motion is multiplied. At sea, it is applicable only to the screw, which revolves from 50 to 80 and even 100 times per minute. Formerly, a speed of piston to produce this rapidity of revolution directly, was considered impracticable, even with very short strokes, and it was universally the custom to obtain it by the intervention of multiplying cog wheels, with one set of cogs made of oak set in iron wheels. But the screw engine, having direct action, with a stroke of 3 and 4 feet and 80 revolutions, has been brought to a speed of piston reaching 600 feet per minute. There are, however, engineers who regard gearing as preferable to this, and symptoms are not wanting of a tendency back again to its use. It strains the ship and engine less.

18. The "Screw Engine" is, of those forms enumerated, confined to the geared, the horizontal direct or back action, the steeple, the trunk, and the oscillating (as the English corvette Esk) forms.

19. Generally, in judging the various engines proposed or in use for marine purposes, either as respects form, or detail of construction, the questions which arise refer to the space occupied; to accessibility of parts for cleaning, oiling, or repairing; to light and air in the engine room; to protection in men-of-war against enemy's shot; to loss of heat by radiation; to the means of packing against escape of steam; to the degree of friction, or jerking tendency; to any unusual liability in the engine to be thrown out of line; to provisions against damage, by elongation of the pipes or otherwise, from expansion; to lubricating arrangements; to economy of power—by the vacuum which can be produced, by the valve and other openings for

the free admission and egression of steam, and by devices for controlling its expansion in the cylinder; to the exertion of force which does not act propulsively, for that which does not, costs like that which does; to either a want, or an excess, in the size, strength, weight, proportions or number of the parts which compose the engine; and to the pounds of coals per horse power per hour needed to run it.

HIGH AND LOW PRESSURE.

20. When an engine is so arranged that steam from the cylinder escapes into the air, such engine is a *non-condensing*, or, as it is more commonly termed, a *high pressure* engine. And when arranged so that the steam escapes into a condenser, as described in Art. 2, the engine is a *condensing*, or, as more commonly termed, a *low pressure* engine, because by producing a vacuum in the end of the cylinder open to the condenser, not only the *excess* of pressure in the boiler (see page 16, et seq.), but the atmosphere of steam which counterbalances the external air, is rendered available as an effective force on the piston, and admits of *lower steam*. Consequently, when the low pressure engine is working steam of 10 pounds pressure per square inch by the steam gauge, the effective pressure per square inch exerted on the piston is, or may be, 20 or 25 pounds—viz., pressure per gauge, plus the vacuum, which is, however, rarely perfect, or more than about 12 pounds.

But without the condensing apparatus attached, the engine is a non-condensing, or high pressure engine, in which the exhaust pipe m, (Fig. p. 53,) would lead and deliver its steam into the air. Consequently, instead of a vacuum, as in Art. 2, a force equal to the atmosphere itself would always exist in the cylinder to oppose the motion of the piston; and hence, the *effective* pressure upon the piston would equal only the excess of steam in the boiler indicated by the safety valve, (Art. 25, Sec. 1,) and require *higher steam.*

The *high pressure engine*, and its relative advantages or disadvantages in different situations as compared with the low pressure, will be more fully explained in the succeeding Section.

SURFACE CONDENSATION.

21. The Fig. page 53, exhibits at *o*, the injection water which condenses the steam and mingles with it in the condenser *n*, from whence it goes into the reservoir *v*, and thence, so much of it as is needed, into the boiler, salt and all, as feed, rendering it necessary to blow occasionally, (Art. 41, Sec. 1.) But this evil is now obviated by the " surface condenser," in which the injection water is showered down upon and amongst a multiplicity of small tubes, through which the exhaust steam is passed and condensed, without mingling with the condensing water. Consequently, by theory at least, the condensed steam, being exactly equal to the evaporation, and all of it returned to the boiler, no sea water need ever go into the boiler to supply evaporation, nor any blowing ever be necessary. But practically, there will be some leakage of the steam to be supplied from the sea, and some sediment to be blown off, both which will occasion a small demand for sea water in the feed.

22. The Fig. on the opposite page exhibits the parts of a surface condenser, and the manner of its operation.

A A is the condenser; *b* is the exhaust pipe from the cylinder, and corresponds with *m* in p. 53; *l* is the injection, which corresponds with *o* in p. 53. Now if all the pipes and partitions seen in the condenser A A

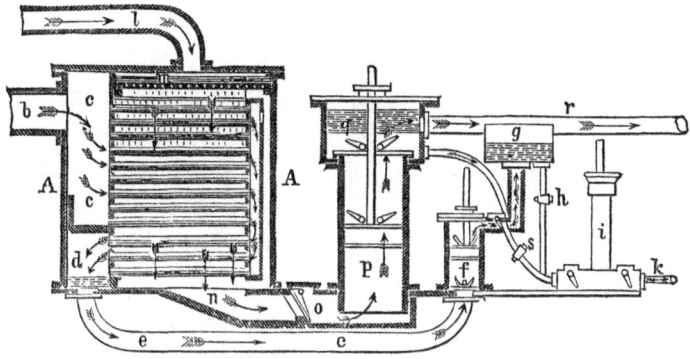

were removed, it would become like the condenser *n* p. 53, and the exhaust steam and condensing water would mingle as they do in p. 53. And when these tubes, etc., become leaky, as is sometimes the case, rip them out, and A A becomes no longer a surface condenser, but operates precisely in principle like that in p. 53.

23. The operation in A A, having all its tubes and partitions perfect, is thus. The exhaust steam *b* enters an upper chamber *c*, thence passes and repasses, as the arrows show, through small tubes into a lower chamber *d*, where, its heat having been extracted by the condensing water showered upon and among the tubes from the injection *l* and strainer *m*, the steam has become condensed to water.

This flow of steam through the pipes, and shower of water upon them, are both promoted by artificial means—the former by the *fresh water air pump* f, which draws through the *fresh water channel* e e ; and the latter by the *salt water air pump* p, which draws through the *channel* n *and foot valve* o.

The fresh water air pump forces the water of the condensed steam into the *fresh water reservoir* g,

66   THE STEAM ENGINE.

whence it is drawn for feed; and the salt water air pump forces the condensing water into the *salt water hot well* q, whence it mostly goes into the sea through the *waste pipe* r.

The force pump *i*, draws fresh water through the *fresh water feed cock* h, and forces it into the boiler by the feed pipe *k*. Any deficiency of fresh feed water, is supplied by opening the *salt water feed cock* s, whence the pump *i* draws so much as may be found necessary, through the curved pipe which appears in the Figure.*

24. The quantity of condensing water, and water of condensation, are the same with both condensers, the old one page 53, and the new one represented by this Figure. The work performed by the single air pump of the former, is therefore the same as that performed by the two air pumps of the latter. And the feed water in both cases has an equal temperature, 100° or 105°. So that the gain with a surface condenser is reduced to that due to not blowing, and to the cleaner condition of the boilers—which latter is marked, because a deposit of the lime in water the boilers contain, once made, it can cause no more scale; whereas when the water is changed, every new supply adds to the scale, by a new deposit of lime, if not of salt.

HEATERS FOR FEED WATER.

25. Of course when the steam and condensing water mingle, they immediately assimilate in temperature, as parts of the same body of water. So it appears, that in the surface condenser, where the two do not mingle, they nevertheless assimilate, and stand in

---

* The apparent break in that pipe is not intended, but is an error of the artist.

their respective hot wells at 100° or 105°, about the temperature at which feed water usually reaches the boiler, provided it is not raised in temperature during its passage from the hot well.

26. A very common practice now is, to run the feed pipes through a "heater," in which the feed water is raised to 135°. Most generally this additional heat is derived from the blow water, where there is any; and all the heat the feed abstracts from it, is of course so much saved—a clear gain. But this source of heat to feed water is wanting where the surface condenser is used; and in such cases the heat has been derived from a water jacket about the smoke stack, above the steam chimney—which arrangement, with the circulation in the feed pipe leading to and from it shut off through forgetfulness, caused the explosion on board the Great Eastern. In some cases the heat has been derived from a feed pipe run up and down spirally, a short distance, within the chimney, to save heat and turn it to account (Art. 39, p. 26).

The cost, the complication, and the danger arising from these arrangements constitute an objection.

EXPANSIVE ACTION OF STEAM.

27. Before Mr. Watt's time, steam, when applied to engines, was used merely as an agent to produce a vacuum, and the weight of the natural atmosphere acting on the vacuum (from 10 to 14 pounds per square inch according to its perfection), was the *real* reliance for power. And although such engines were sometimes called steam engines, they were also, and more properly, termed "Atmospheric Engines."

28. They had cylinder, piston, piston rod, valves, and a working beam. One end of the cylinder was

always open to the air, and for the purpose of producing the necessary reciprocating motion, two cylinders were needed, one acting on each end of the beam; or else, a single cylinder to work one end of the beam, and a heavy weight to work the other: that is to say, one end was worked by the gravity of the air, and the other end by the gravity of a solid. See Art. 1, page 9.*

29. To condense the steam and produce the vacuum (see foot note page 54), a cold bath was administered to the exterior of the cylinder, which plan of condensation left the cylinder so cold, that it condensed all the first steam admitted of the succeeding charge.

* The atmospheric engine here considered, which works by the weight of the air, is not to be confounded with Ericsson's hot air engine, which is driven principally by the expansive force of heated air. The prime mover of the one is gravity, of the other in the main elasticity. See Art. 1, page 9. The bottom of the cylinder is the top or side of the furnace. The opposite end of the cylinder is open, and the reciprocal motion is produced by two cylinders as with the atmospheric engine; or else, as in it, with one cylinder, and a wheel loaded on one side—an equivalent for a weight on the opposite end of a beam—in which case the engine works in part by elasticity and in part by gravity.

To work the engine, cold air is admitted beneath the piston, there takes heat from the furnace, expands and drives the piston up, or outward, and by a simple but ingenious arrangement is then expelled from the cylinder to permit the downward or inward movement of the piston, which is probably also aided in this movement by the weight of atmosphere acting with a partial vacuum beneath the piston. So far as that *is* the case, the driving force is further compounded of weight and elasticity.

It is intended at the end of the book, to enter minutely into a description of this engine, and the manner of handling it, because of the important part, in the opinion of the author, it is destined to play, even on board sailing men-of-war, in driving a foul air force pump, and banishing the frightful epidemic which so often decimates their crews. Placed in the hold of a ship and geared to a pump, or to a much better valve arrangement which has been or may be proposed, or to a fan blower, it will effect its purpose by a double operation, similar to that by which a steam engine frees a ship from water by the force pump and bilge injection acting unitedly. See second foot note, page 57.

The *caloric* engine is compact, simple, perfectly safe, exempt in fact from explosion; is worked by a person of ordinary intelligence though not a mechanic of any sort; and if of only two horses power, consumes less fuel than an ordinary small sized parlor grate. A system of fixed and permanent exhaust pipes, ramifying through a ship, and a shifting hose to lead in the magazine occasionally or other parts at pleasure, with this engine attached, would make a very effective combination.

THE STEAM ENGINE. 69

30. Mr. Watt gained the reciprocating motion with a single cylinder closed at both ends, and with the elastic force of steam as his sole agent, dispensing entirely with gravity (see again Art. 1, p. 9). He made the steam, which before had performed the single office of producing a vacuum, now perform the double office of producing the vacuum, and moving the piston with its elastic force.*

31. He introduced the separate condenser, with a continuous jet, by which the cylinder always remained hot, and the condenser cool;—or reversing the expression, the cylinder was never cool, and the condenser never hot—two very important considerations in working a steam engine. The air pump was a necessary consequence of the condenser and jet.

32. But Mr. Watt's engine, as worked in his day, and it may be said until a comparatively recent period, though strictly a steam engine, possessed in principle but little more power than the atmospheric engine; but yet, owing to greater perfection of detail, its action was much more efficient. He had mainly substituted the elastic force of an atmosphere of steam for the mechanical weight of the natural atmosphere (Art. 1, page 9). But as 4 or 5 pounds of steam by the gauge, over the atmosphere, was needed to start the engine from a state of rest, it was used; and as a liability to stop is ever imminent, an excess of pressure, at least 4 or 5 pounds, was always carried.† And in

---

* It is difficult to find an expression to suit this idea. If steam acts on one side of a piston on the other side of which a vacuum exists, it is usual to speak of it as moving *against* a vacuum. But the word against implies opposition, whereas a vacuum is aid, not opposition. Hence the expression *with* a vacuum.

† Although a steam vessel will when in motion run a long while "on a vacuum" as it is termed, that is with no steam shown by the steam gauge standing at zero, and engineers often to save labor or fuel when near the end of a trip let the steam go down and run on a vacuum, it is nevertheless a dangerous practice, and ought never to be allowed, for the officer in command should always rely upon ability

working the engine, not only the exhaust valve remained open to the condenser, but the steam valve remained open to the steam chest and boiler, during the whole stroke of the piston from end to end of the cylinder. By this, the steam followed the piston and filled the cylinder entirely with it, of a uniform elasticity, about equal to that in the boiler.

33. Now, however, whilst Mr. Watt's principles are all retained, and also his detail for the most part, the practice of working low steam of 4 or 5 pounds is entirely exploded, and few condensing engines are run with less than from 15 to 25 pounds in the boiler; though in working the steam into the cylinder, it is not allowed to follow the piston more than a fraction, say a fourth, or half the length of the cylinder, when the steam valve closes, and the piston is driven through the remainder of its stroke by the expansive force of the high steam admitted into the cylinder before the valve closed. This is what is meant by "using steam expansively," or "the expansive action of steam." Its effect is—without increasing the cost of boilers because it admits a reduction of steam room, whilst it requires an increase of the strength and weight of the parts of an engine—to reduce very largely the consumption of fuel, as compared with the force exerted. To explain this may not be difficult.

34. In Art. 8, page 34, it is said, high steam is merely low steam forced and confined in a smaller compass, by which compression the temperature is increased—that is, sensible heat is developed from the latent state; the volume (Art. 48, p. 29) is decreased according to the table, not exactly yet nearly in proportion to the

---

to back the engine suddenly at any moment, which cannot be done when it runs on a vacuum. So it has been proposed to fight steamers on a vacuum, by which, in case of perforation of boilers by a shot, air might enter instead of steam escaping. But this is absurd, because the ability to start, stop and back must be preserved.

reduction of bulk; and the pressure increased in proportion, very nearly, to the decrease of bulk. And vice versa, low steam is merely high steam expanded, by which its temperature, volume, and pressure, are affected, according to the degrees of expansion.

35. Thus for example, if a cylinder full of steam having 5 pounds pressure above the atmosphere, is compressed into half the cylinder, it becomes steam of 25 pounds pressure above the atmosphere. And, on the contrary, if steam of 25 pounds above the atmosphere be admitted to fill only half the cylinder, when expanded to fill the whole cylinder it becomes steam of 5 pounds above the atmosphere.

36. In the one case, 5 pounds of steam per gauge is, with the atmosphere of steam in the boiler added, effectively 20 pounds in the cylinder of a condensing engine, by theory.

37. In the other case, 25 pounds of steam per gauge is, with the atmosphere of steam in the boiler added, effectively 40 pounds in the cylinder of a condensing engine, by the theory.

38. Therefore, compress a cylinder full of the steam at 20 lbs. effective, volume 1281, and temperature 228°, into half its bulk, and it becomes steam of 40 lbs., volume 670, and temperature 269°. Re-expand it again to fill the cylinder, and it again resumes its original pressure, volume, and temperature. Practically, the water it contains remains nearly the same in both cases; and the sum of the latent and sensible heat remains the same. Hence, the fuel expended to produce the whole cylinder full of the low steam, is the same as that needed to produce the half cylinder of high steam.

39. But if fuel is applied to make steam of 40 lbs., instead of applying it to make steam of 20 lbs. pressure, and if half a cylinder full of the former is used

and expanded, instead of a whole cylinder full of the latter, there will be a gain, usually made apparent and estimated as follows:

40. In the first of the two cases, Art. 39, the pressure, including the vacuum, is, in the cylinder at the beginning of the stroke, 40 lbs., at the middle it is 40, and at the end it is 20 pounds, making an average of $(40+40+20) \div 3 = 33\frac{1}{3}$; whereas in the second case, the pressure is 20 at the beginning of the stroke, 20 at the middle, and 20 at the end, giving an average of 20 pounds. This exhibits a gain of power exceeding 50 per cent., free of cost in fuel or in any other way. What, under certain drawbacks, the gain is in practice, is a question among engineers. They have a nicer way of calculating the theoretic gain by using tables of hyperbolic logarithms and a rule, both which may be found in Haswell, page 209.*

41. The only objection urged against working expansively and cutting off short is, that the high steam it requires produces great loss of heat by *radiation*, and heats ships' fire rooms below decks almost beyond endurance, unless well ventilated. Radiation is in a measure prevented by jacketing the boilers, pipes and cylinders with felt cloth—the latter even with steam.

---

* As an example, to compare the results of the rough and the nicer methods, suppose steam admitted to the cylinder at 40 lbs. effective, and cut off at $\frac{1}{4}$ the stroke. Then by the rough method the pressures at the several points of the cylinder will be, at the beginning 40 lbs.; end of the 1st quarter 40 lbs.; end of the 2d quarter 20 lbs., the steam being expanded into double the space; end of 3d quarter, steam at 20 lbs. being expanded into a space $\frac{1}{2}$ larger, is $\frac{1}{3}$ less in force, or 13.34 lbs.; at the end of the stroke, the half cylinder full at 20 lbs. having become expanded into double its space, is reduced to $\frac{1}{2}$ of 20, or to 10 lbs. The average pressure is, therefore, $(40+40+20+13.34+10) \div 5 = 24.66$, a gain of 100 per cent. or over, which arises from cutting off shorter, and all without cost.

In Haswell, page 209, the same example worked out by hyperbolic logarithms gives 23.87 lbs. as the result.

SECTION VI.

The High Pressure Engine. Relative advantages of the High and Low Pressure Engines. Locomotive Engines Power of Engines. Calculated and indicated horse power compared. Consumption of fuel per horse power. Getting up steam and managing it and the engines. Link motion.

THE HIGH PRESSURE ENGINE.

1. The high pressure, more properly termed the non-condensing engine, invented by Oliver Evans of Philadelphia about the year 1780, differs from the low pressure or condensing engine invented a few years earlier by James Watt, chiefly, as before remarked, in dispensing with the condenser and air pump, and leading the exhaust or eduction pipe into the open air, instead of to the condenser.* The resistance which the exhaust steam meets in the air, reacts upon the piston, and to the extent of that reaction opposes and resists the motion of the piston, requiring higher steam in the boiler to produce a given effective pressure on the piston.

RELATIVE ADVANTAGES OF THE HIGH AND THE LOW PRESSURE ENGINES.

2. The principal advantages of the high pressure engine, are, its reduced weight, reduced room occupied, and reduced cost of construction. And these advantages arise chiefly from dispensing with the air

---

* If an opening is made in the channel between the condenser and air pump, by taking off the bonnet over the foot valve k (Art. 4, p. 57), a condensing is converted into a non-condensing engine. This expedient is often resorted to when the condensing apparatus becomes deranged. A wooden box pipe leads off the steam.

pump, exhaust pipes, condenser, and the larger cylinder, steam pipes and chests, all of which low steam requires. They are all, of course, best done without when possible, because they cost money, weigh heavily, and occupy space, or make friction.

3. The high pressure arrangement, as compared with the low, is more expensive on the score of fuel, though not to the extent which is often supposed. The causes which operate in the high pressure engine, and not in the low pressure, to occasion consumption of fuel are, 1st, the atmosphere of steam to be created for overcoming resistance of the atmosphere in the cylinder, produced by opening the exhaust into the air; 2d, the greater loss of heat by radiation; and 3d, the greater escape of heat by the chimney. On the other hand, the causes which operate in the low pressure and not in the high pressure to occasion consumption of fuel, are, 1st, the power necessary to work the air pump; 2d, that expended to overcome the greater friction of the condensing engine, arising from the greater number of friction points and parts, and the greater weight of those parts; and 3d, the low temperature of the feed water of the condensing engine, as compared usually with the temperature of the feed water of the non-condensing engine.* Deducting the latter from the former set of causes, the consumption of fuel remains probably somewhat greatest in the high pressure engine.

4. The high pressure arrangement, then, being lighter, more simple, more compact, and less expensive

---

* The temperature of the feed water in non-condensing engines (except locomotives) is raised to about 200°, by passing the eduction pipe through the hot water cistern. In the condensing engine, the feed water is generally at about 100°, unless raised by the heater (Art. 25, page 67), which reduces, though it does not remove this particular cause of difference in the two engines.

in construction, is used on the western rivers,—where the shoal water renders light draft important; where there is no space in the hold below the water line for the condenser, and water for it would have to be raised at a cost of power; where boats so soon decay, or are destroyed, as not to warrant the expense of condensing engines; where the turbid waters would choke the condenser with mud, and wear out the foot valve and air pump cylinder; and where capital to build expensive engines is comparatively scarce, and fuel cheap. On the other hand, at the east, where draft of water is of less consequence; where boats carry less freight and last much longer; where weight, and expense of construction, are of less importance than the saving of fuel, and there is room below for a condenser; and where the popular feeling is altogether, though unreasonably, averse to high pressure engines as more dangerous in boats; all of them are found to have condensing engines.

5. In marine steamers, the high pressure engine is desirable on account of economy of room and weight, though objectionable because of the perhaps somewhat greater consumption of fuel, but more for the radiation from 140 pounds of steam of 360° temperature, which renders the fire room, and indeed the whole ship below intolerably hot.

6. In land engines, where there is room to set boilers of great length, which in consequence abstract from the flues all the heat that is available for making steam before they connect with the chimney, and where also the boilers are bricked in so as in a great measure to prevent radiation, two of the enumerated causes of increased consumption of fuel with the high pressure engine do not exist. Hence, for stationary land engines, the high pressure is asserted to be most economi-

cal, and therefore generally used. It is also less complicated, requiring less skill to manage, and less cost to keep it in repair. Water for the condenser of a low pressure engine on shore, costs money if obtained from hydrants, and costs power if obtained from wells; which, and sewering off the waste water, constitute direct arguments against the low pressure and in favor of the high pressure as a land engine.

LOCOMOTIVE ENGINES.

7. Locomotive engines, which must be light and compact, and use the least possible supply of water, are necessarily non-condensing. The locomotive differs from other non-condensing engines usually seen on shore, in having neither a *balance wheel* nor a *governor*,* and in the arrangement of the steam pipe, which takes the steam to the cylinders.

This steam pipe in a locomotive engine is not apparent, but is wholly within the boiler and smoke box. It extends from near the top of the dome, through the whole length of the steam room, and through the back end of the boiler, into the smoke box, where it branches, and carries steam to each of the cylinders. These cylinders may usually be seen, in locomotives, secured, one on each side. The dome at the forward end of the boilers is placed over the furnace end of the boiler, because it makes a heavy part of the boiler, and the greatest weight must be near the driving

---

* The object of both these appendages (the balance wheel and governor) to an engine, is to equalize its motion. Another principal use of the balance or fly wheel is to take the engine past the centres, or dead points. Marine engines and locomotives do not require either of these appendages, because in them uniformity of motion is not important, and the impetus of the vessel or engine, carries the crank past the dead points. And what makes the paddle wheel in some degree more a balance wheel, it has one or more iron buckets or paddles set opposite the crank as a counter balance. All cranks in all engines are in some way counterbalanced.

wheels, in order to produce a friction which shall cause them to adhere to the rails.* The driving wheels, and therefore the dome, must be at the opposite end of the boiler from the cylinders; hence the steam pipe must run the whole length of the boiler, and runs within instead of without the boiler, in order to prevent the loss of heat from that pipe by radiation, which would be especially great in the rapid currents of cold air to which locomotives are peculiarly exposed.

CALCULATED AND INDICATED POWER OF ENGINES.

8. The working capabilities of steam engines are estimated and compared by their *horse power*—that is, if an engine will, in a given time, perform the work which 100 horses ought in the same time to perform, the engine is said to have 100 horse power.† The steam engine was first used at English mines for the purpose of raising water, and other substances. This had been done by horses. Hence the comparison, and the establishment of the horse power as a unit.

9. By the ordinary estimate, one horse should raise 33,000 pounds, through one foot, in one minute. Twice that weight through half a foot in a minute, or half that weight through two feet, is the same thing. If the product of the weight in pounds into feet per minute be 33,000, it represents an estimated horse power. *Calculate the pressure on the piston*, by multiplying

---

* Some locomotives have two pairs of driving wheels, by which increased adhesion is obtained without bringing more weight at any one point than the rails can bear without bending A driving wheel of 6 feet diameter revolves 290 times in a mile; or, at 30 miles an hour, 145 times per minute; and with two feet stroke, the piston travels 145 × 4 = 580 feet per minute, or nearly 10 feet per second!

† But six hours a day, one day with another, is all a horse can endure. Therefore, a 10 horse power engine, working constantly, does actually the work and saves the stabling of 40 horses.

its area in square inches into the effective pressure per square inch; multiply the *calculated* pressure thus obtained by the speed of the piston in feet per minute; divide the product by 33,000; deduct $\frac{1}{10}$ for power consumed in working the air pump, and $\frac{2}{10}$ more for friction, and the remainder is the net "calculated horse power" of the engine.*

10. But this *calculated* horse power exceeds the actual horse power at which an engine works, because the pressure in the cylinder which is the true measure of the work done, may never exactly come up to the pressure in the boiler which is that used in the calculation; and the vacuum in the condenser, never perfect (as shown by the vacuum gauge on the condenser), is rarely equalled by that in the cylinder, at least until the latter portion of a piston stroke, because when the exhaust valve opens, the equilibrium is far from being instantaneous.

11. To determine then, how much the actual pressure of steam in the cylinder falls short of the pressure in the boiler as shown by the steam gauge; also how much the vacuum in the cylinder falls short of that in the condenser as shown by the vacuum gauge placed on the condenser; an instrument, called "the indicator," when attached to the cylinder, is so arranged as to

---

* There are complicated rules given in the books for calculating this power with greater accuracy. To compare the results of this rough, with the more accurate calculation by nicer rules, take the example given in Haswell, page 221, (Ed. 1860), in which the steam per gauge is 20 lbs. Add 14.7 lbs. and the effective pressure is 37.4 lbs. The steam is cut off at $\frac{1}{4}$; roughly, the average pressure through the whole stroke is 21 lbs.; by the rule (Ib., p. 209) it is 20.86 lbs. The stroke is 10 feet; revolutions 20; consequently the speed of piston is 400 feet per minute. The diameter of the cylinder is 60 inches, hence the area of piston (Ib., p. 93) is 2827.4. Roughly calculated, therefore, the gross horse power is $(2827.4 \times 21 \times 400) \div 33,000 = 720$. From this deduct about $^2/_{10}$ for friction, and $^1/_{10}$ for power used in working the air pump to produce the vacuum (over $\frac{1}{4}$ the total used—see foot note page 54), and the remainder is the roughly calculated net horse power, which exceeds the nicely calculated only about $^1/_{10}$.

mark, with a pencil, on a paper card, a figure, the lines of which indicate the amount, both of steam and of vacuum, which acts in the cylinder upon the piston at every portion of the stroke.

12. A specimen "indicator card" (here shown) for a half revolution of the engine, gives both the force of steam and of vacuum in a half revolution—though usually an indicator card exhibits the force in an entire revolution—and is read as follows:

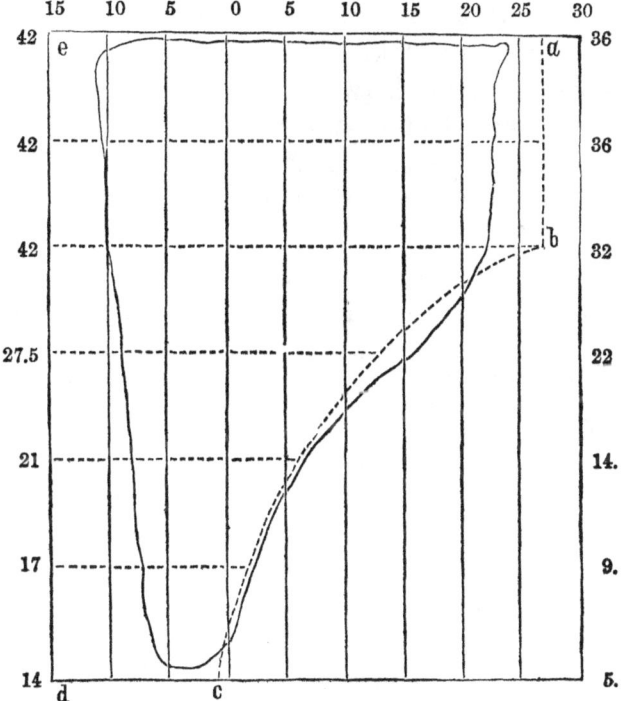

13. The continuous irregular line which bounds the harp-shaped figure, and the zero or vacuum line drawn vertically from 0, are the only ones, of the several lines given in the cut, which are marked by the indicator pencil. The other vertical parallel lines are meridians,

drawn for the purpose of measurement; those to the right of zero for measuring the steam, to as high as 30 pounds if so much is carried above an atmosphere; and those to the left, the vacuum, say 15 pounds or so much as is obtained. The card itself, with the indicator lines only, is copied from one taken on board the Wyoming, and published in the Journal of the Franklin Institute, Vol. 68, page 350; and at the time it was taken, the engines worked steam per gauge 27 lbs.; vacuum 23.5 inches of mercury, or $11\frac{1}{2}$ lbs.; revolutions $80\frac{1}{2}$; stroke 3 feet, cutting off at about $\frac{1}{3}$, perhaps a little over, taking into the account the *clearance*—as the space between the piston and cylinder head at the end of a stroke is called, and amounts usually to a couple of inches.*

14. Now if the steam in the cylinder could equal exactly that in the boiler above an atmosphere, viz., 27 pounds, and the vacuum below be at the same time perfect; and if in following the piston during the first third of the stroke, down to the point of cutting off, the uniformity of pressure is preserved, then the parallelogram bounded by $a\,e$ and $a\,b$ should indicate the pressure that far. And the valve closing at b against a further admission of steam, the hyperbolic dotted curve line $b\,c$ should, with the figure bounded by $b\,c\,d$, indicate the decreasing pressure in the cylinder during the remainder of the stroke. So also should the seven ordinate lines or parallels drawn across the meridians, from $e\,d$ to $a\,b\,c$, measure the effective pressure in the cylinder at each point of the stroke where they are drawn; in which case it would be as registered in the

---

* This *clearance* is reduced as much as possible consistently with safety to the cylinder head, because at every stroke the steam which fills the space is lost. An objection to obtaining rapid revolutions by speed of piston instead of by gearing (see Art. 17, page 62) is, that it occasions greater because more frequent loss of steam in the clearance and steam passages.

left hand column of numbers, viz.: at the beginning of the stroke the steam, measured on the upper line to the right hand of zero, shows as 27 lbs., which, added to 15 pounds for the vacuum, registers 42 lbs.; at the second line 42 lbs.; at the third 42; at the fourth, being $12\frac{1}{2}$ for steam and 15 for vacuum, it registers 27.5 lbs.; at the fifth 21 lbs.; at the sixth 17, and at the seventh 14 lbs.—averaging 29.4 pounds, or, about the same thing as the formula in Haswell (p. 209) would give as the calculated average.

15. Actually, however, the steam on entering the cylinder gives a more rapid motion to the piston than it can follow with full force. Hence the pencil, marks on this indicator card to the right of zero a less pressure than 27 lbs., hardly reaching 23 pounds, though the valve remains open and steam follows the piston to b.

Below b, after the valve has shut, the pencil line is seen in the cut to swell beyond the dotted curve, indicating that the valve has not closed tight, otherwise the line would be within, below b as well as above.

16. The vacuum mark as traced by the pencil, begins at the bottom of the cut, on the vacuum or left hand side of zero, where, at the first moment of opening the exhaust, it registers as 5 lbs. But the longer the exhaust is open the more the vacuum improves, until at $\frac{2}{3}$ the stroke it exceeds 10 lbs., as is shown by the pencil mark crossing the meridian line drawn from the upper graduation 10. At the end it reaches between 11 and 12 pounds, equal to the maximum which exists in the condenser (Art. 13). So great a difference between the vacuum at the beginning and at the end of the stroke, is not, however, usually exhibited by indicator cards.

17. The sums of the actual pressures by steam and by vacuum, or the indicated effective pressures, at the

several points, as measured on those parts of the ordinate lines included within the figure traced by the indicator pencil, are set down in the column of numbers to the right hand of the Figure page 79; and these, added together and divided by 7, give 22 pounds as "the average indicated effective pressure."

18. When the horse power is computed from the calculated average effective pressure (Art. 14), the result obtained is the calculated horse power. And when the result obtained is from the indicated average effective pressure (Art. 17), it is the indicated horse power. In the particular case adduced, the Wyoming's, the difference between the calculated and indicated horse power is upwards of 15 per cent.; the former being greatest, and the latter most correct.

CONSUMPTION OF FUEL AND "MARINE ECONOMY."

19. When boilers are constructed of such size and capacity for generating steam, that they can, by urging and stirring the fires, be made to produce much more than is ordinarily wanted, engineers find great economy of fuel by spreading the coal in a thick bed upon the grates, and permitting it to consume quietly—for then the consumption is more nearly perfect, and less is lost in smoke, or by admission of cold air at the doors when they are opened to stir and force the fires, or by other air which passes up through cracks made in the coke bed by breaking it up, which air also passes imperfectly heated, and but partially deoxygenated, into the flues with damaging effect upon the generation of steam.

20. It was formerly usual, when the arrangements of a boiler and engine were ordinarily complete, to estimate the consumption of coals at about 10 pounds

per horse power per hour, which would now be considered enormous, and the constant aim of both constructing and running engineers to reduce this rate, has been so far successful, that 5 or 6 pounds is now not unusual. The next Art. will show much better results.

21. The case of the Wyoming, cited in the foot note to page 30, exhibits a remarkable success, in which the consumption was reduced to 2.73 lbs. of coal per hour per horse power. This result must be due principally to the perfection of the boilers and their freedom from scale, as well as cleanliness of the flues from soot, to their correct proportion as relates to the engines, to quality of the fuel, and to judicious firing; and due subordinately to the excellence of form, construction and adaptation of the engines. Yet the agency of the engines in the result, must have been mainly negative—that is, not in creating or developing, but only in saving power against loss by radiation, leakage and friction; because the engine affords a positive gain in no way, except by the condenser, and by expansion of the steam in the cylinder— if, indeed, they can properly be styled gains.

22. The Great Eastern is said to have reduced the consumption to about 4 lbs. per hour per horse power; locomotives to $2\frac{1}{4}$ and $2\frac{1}{2}$ lbs.; some marine engines the same; Cornish engines to 2 lbs.; and the bold prediction is in print, that the consumption will yet be reduced to 1 pound per hour per horse power!

23. The Wyoming's total consumption per hour was 2,948 pounds, per day 31 tons,* horse power ex-

---

* Art. 1, Sec. 8 of the Constitution of the United States empowers Congress to fix the standard of weights and measures; under which 2,240 pounds is a ton; and when a ton weight is spoken of in ships, it is understood to be always of 2,240 pounds, or a long ton. In many places, by local municipal regulation or custom, a ton is 2,000 pounds, or a short ton. But even such cases have not stood the test of a legal appeal.

erted 1088, speed 11 knots. At these rates, carrying 266 tons of coal, her supply would last $8\frac{1}{2}$ days full steaming, and drive her 2,240 miles, assuming the wind and sea to at least compensate, one day with another, in the variables.

24. The law of resistance is, that it increases or decreases with the square of the velocity; and so with the fuel used to overcome the resistance. Hence, if it be desired to know what consumption per hour will drive the Wyoming 8 knots an hour, or 200 miles a day, make the statement, as $11^2=121$ is to 2948 (pounds per hour to produce 11 knots), so is $8^2=64$ to 1560 pounds per hour, or 16 tons per day, which should drive her 8 knots per hour, 200 knots per day, or, with 266 tons of coal, 16 days and 3,200 nautical miles.

For the third term of the foregoing rule of three proportion, substitute 6^2, and the rate of consumption to give 150 miles a day may be worked out, and found reduced to about 800 pounds per hour, or to 9 tons per day, and the time and distance increased to 19 days and over 4,000 sea miles. And so on, the lower the speed, the greater the distance which can in theory be accomplished with a given supply of fuel on board.

25. Practically, there is a limit, both in speed and in consumption, below which the gain ceases; and this is owing principally to the loss of heat by radiation, which though it may be small, if long enough continued, even without steam enough to produce any motion, would in time cost the whole of the fuel, just to supply the radiation. With a sea or swell on, the reduction of speed is less favorable, because it is accompanied by a loss of the momentum necessary to go through or over the waves.

26. Nor, on the other hand, is it wise to rack a vessel and her machinery, and waste fuel, by butting her

square into a heavy head sea; or to keep the course and stem the wind, when by keeping off a little, the resistance often greatly decreases, the fore and aft sails draw with very marked advantage, and the speed so far increases, that although steering angularly with the true direction, the distance from the point of destination reduces more rapidly than it would on a direct course, but with less speed.

27. In certain latitudes and seasons, a long passage exposes a ship more to gales, to hold on through one of which might cost more fuel than a continued economy had saved. Practically, therefore, it is for many reasons possible to go too slow as well as too fast for economy. In this, experience and judgment must govern, for no fixed rules are applicable under all the varying circumstances.

28. With currents, the case is different. It is best to meet them directly, unless an eddy offers; and the rule for speed in steering against them with reference to the best economy of fuel, is to let the speed through the water be once and a half the current; that is, if the current is 4 knots, let the speed per log be 6 knots. The books demonstrate this by figures.

29. In running with steam as auxiliary to canvas, when there are several boilers, it is common to shut off one or more; or to reduce consumption by throttling off close; or by cutting off very short and expanding the steam very much—for which reason, and to accommodate the action of the engine to increasing or diminishing breezes, adjustable cut offs are introduced, which can be altered whilst the engine is in motion.

30. When it is desired to run solely with canvas, yet keep the water hot so as to be able to steam up on short notice, the fires are banked in the furnaces, and

then care is only necessary to "keep a vacuum out of the boilers"—that is, to keep an atmosphere of steam in the boilers. The water gauge cocks will show when this is not the case, for if not, when opened air sucks in. The evil of this vacuum in the boilers is, that it makes the lap joints of the boiler plates leaky, by bringing upon them an unusual and reverse pressure, which breaks the rust.

GETTING UP STEAM AND MANAGING IT AND THE ENGINES.

31. When an engineer is directed to get up steam preparatory to moving, his boiler being clean and tight, he first fills it with water to the upper cock, that he may have water enough to allow liberally for evaporation whilst making steam, and before the feed pumps act. If the blow off pipe leads through the bottom of the vessel, and the boiler is in the hold, opening the cock of that pipe, and raising the safety valve, will nearly or quite fill the boiler;—otherwise it must be entirely filled by the hand force pumps with which boilers are provided.*

32. He next lights his fires, closing the damper so that they shall kindle, and the heat increase, slowly, the object being not suddenly and irregularly to expand the parts of a boiler; for if the flues heat, and elongate much by expansion before the shell (especially the legs) also heats, a strain and leak somewhere will be likely to occur. Hence the moderation exhibited by

* Donkey engines (with a small separate boiler) are common in lieu of hand power to work this pump. The beauty of one is, that it is available for pumping in feed against high pressure, which hand power cannot do; and is specially valuable for keeping boilers supplied when it is inconvenient, or owing to any circumstance impossible, to work the engine and the main force pump *o* (Art. 4, page 57).

engineers in getting up steam, and it is usually commendable.*

33. The safety valve is kept open during the process of raising steam, for the escape of rarefied air, which, if confined in the boiler, would increase pressure on the surface of the water, and not only retard the formation of steam, but prevent protection where the flue emerges from the water and passes through the steam. So soon, however, as the emission of steam from the safety valve is so great as to produce a blowing noise, the valve is closed, for then the air is out, *an atmosphere of steam having taken its place.*

34. When, in a condensing engine, the steam gauge shows a few pounds excess of steam over the atmosphere, the engineer raises the valves of the engine, and lets steam flow through the cylinder and all the pipes, which opens the *blow-through valve w* that some engines have as in the figure p. 53 near the bottom of the condenser. This operation expels the air, and warms the cylinder, the pipes, &c.; so that they will not condense the working steam when the engine is set in operation.† Warming these pipes expands them; but to prevent their straining by expansion and contraction, they are provided with "slip joints," as may be seen, or an equivalent arrangement on all the pipes, especially the main steam pipe, subject to expansion or contraction under temperatures varying considerably, or from working in the frame of the ship.

---

* There are often emergencies in which a commander, especially in military service, must have steam in the shortest time compatible with safety to the boilers, and even at some risk; and it is well he should know the limit, else it may be in the power of others to balk an enterprise the execution of which should lie solely in his own hands. See Art. 39, et seq.

† The better and more usual plan is to dispense with the valve *w*, and by keeping the exhaust valves closed and blocking open both steam valves, admit steam to the cylinder to warm it and the steam pipes without warming the condenser, which is to be avoided (Art. 31, page 69).

35. When at least 4 or 5 pounds of steam are shown by the gauge, enough to work the air pump and produce a vacuum, the engine may be reported ready to move. The order to start being given, the engineer opens the injection cock a little, works the steam and exhaust valves by the starting bar shipped in the rock shaft, which produces reciprocating motion in the piston, and revolution in the wheels. And when the order to "hook on" is given, he hooks the eccentric rod to the eccentric pin on the arm of the rock shaft, unships the starting bar, opens the injections more,* regulates the throttle valve and cut offs, and the engine continues in motion. He opens also the damper in the smoke pipe, so that the fires may

---

* The engineer opens the injection cocks, in proportion to the quantity and temperature of the steam used, and until the injection water flowing in gives such a vacuum as the engine is capable of producing. He judges of this by a vacuum gauge if he has one; if not, by the sudden ringing noise produced by the exhaust.

If the steam is very low, and the condensing water at a temperature near the freezing point, a quantity of it equal to $5\frac{1}{4}$ times the water evaporated will condense the steam formed, (Art. 6, page 34); but then the water of condensation boils, and the condensation is not sudden or perfect. If the condensing water is of a higher temperature, as sea water generally is, (and it is sometimes as high as 80° in the gulf stream,) the proportion of condensing water must be increased; and if it is desired to make the vacuum instantaneous, the condensing water must be still further increased, and may in some cases reach a quantity equal to 70 times the water evaporated.

To condense steam perfectly and suddenly, the condensing water is admitted by the injection cock, in quantities so great in proportion to the steam to be condensed, that the water will not be raised by the condensed steam to a temperature much exceeding 100°. On the other hand, to condense steam imperfectly and slowly, a less quantity of condensing water will be raised to a higher temperature, even to the boiling point (page 34).

In the one case, feed water from the hot well goes into the boiler at only 100°, and is to be raised to the boiling point by cost of fuel; in the other case, it goes into the boiler nearly at the boiling point.

In the first case there is a saving of power by means of the more perfect vacuum produced, but at a cost of fuel, and of heavier work for the air pump in freeing the condenser; in the second case there is a direct loss of power by the imperfect vacuum, but a saving of fuel, and lighter work for the air pump.

It is argued that upon this principle the vacuum may be too perfect for economy, and cost more in fuel and labor of the air pump than it will come to in power.

be sufficiently active to supply the increased demand for steam. He also regulates the feed and blow off, so as to preserve a good supply of water, and prevent incrustations.

36. When ordered to stop the engine, an engineer first throttles off the steam to check motion, diminishing the injection at the same time, unhooks the eccentric handle from the rock shaft, (which occasions the valves to cease working,) shuts the injection cocks, closes the damper in the chimney, opens the furnace doors, and lifts the safety valve more or less, for reasons mentioned in Arts. 34 and 37, Sec. 1.

37. To back an engine, there being generally but one eccentric to work the steam valves, and that when going ahead, the starting bar is used to work the valves by hand.

38. Most explosions of steam boilers in boats occur immediately on starting from a temporary stopping place, and are consequent on neglect of necessary precautions. Whenever the water has, by this neglect, or by obstruction of the feed pump, as sometimes happens, fallen below the lower cock, (or disappeared from the glass gauge,) it cannot be positively known whether the flues are exposed or not, because the lower gauge cock is usually a little above the level of the top of the flues, and it is then hazardous to open the feed. The only safe course under such circumstances, is immediately to put out the fires in the furnaces. Engineers often prefer to take the risk of explosion, rather than expose their neglect, as they must do by putting out the fires. Such conduct is in the highest degree reprehensible.*

---

* So is reading, conversing, or any other direction of the mind from what should wholly and actively engross the attention of the engineer of the watch whilst on duty.

39. Steam has been raised in marine boilers within 45 minutes, but is not usually in less than an hour and fifteen, thirty or forty-five minutes. And when the air is very damp, the moisture it is loaded with being imparted to the fire, checks combustion, in some instances surprisingly. In damp weather, therefore, there can, with ordinary fuel, be no danger of damaging the boilers by any possible amount of hurry to the fires, though there may be in a peculiarly dry state of the air, as shown by the hygrometer.

40. To facilitate the raising of steam when emergencies are likely to arise, it is common to "bank the fires," just to keep the water hot. In such cases the loss of heat is only by radiation, and it is simply necessary to observe care in keeping "the vacuum out of the boiler"—that is, preventing the external exceeding the internal pressure, which is known by the steam gauge rod, when of the proper length, sinking below zero of the scale. The evil arises from radiation of heat being in excess of the communication.

41. Another good often derived incidentally from keeping up heat in the boilers, is the distillation of fresh water for use in men-of-war. But that of course involves a larger consumption of fuel, because in addition to radiation, evaporation has to be supplied.

DISTILLATION.

42. The best temperature to get the greatest and purest evaporation from a given quantity of coals consumed, is under the boiling point. Extraordinarily, 10 lbs. of water may be obtained from one pound of coals consumed. The weight of a gallon of water (8 pounds) in coals, will produce therefore 80 pounds of water, or 10 gallons. Boiling gives salt in the *spray*.

43. The space occupied by 8 pounds of coals is 276 cubic inches. That occupied by the 10 gallons of water produced, is 2,210 cubic inches. The saving of space by carrying the coal to make water, instead of the water, is therefore upwards of 80 per cent.; whilst the saving in weight is 90 per cent.

44. A common 500 gallon water tank, which holds in weight of water about 2 tons, will contain about 3,200 pounds of coal, sufficient to distill 32,000 pounds, or 4,000 gallons, or 8 tanks of 500 gallons. And whilst no one would think of dispensing with tanks, to hold water when coal cannot be had, it appears conducive to economy of space and efficiency, to fill the tanks with coal instead of water, provided there are no practical objections in the way which a chemical investigation would detect.* In these suggestions there is no novelty.

45. Again, inasmuch as a cubic foot of coal weighs only 50 pounds, whilst that bulk of water weighs $62\frac{1}{2}$ pounds, it follows that if all a frigate's water tanks were filled with coal instead of water, she would be burthened with 20 per cent. less weight, and in effect have 8 times more water.

LINK MOTION.

46. Locomotives, and screw engines, have, on the same main shaft, two eccentrics and eccentric rods for working the slide valve; which valve is the same as, or similar to, that represented in the Fig. page 56. One of these eccentrics is for the forward or progressive, the other for the backward or retrogressive motion of the engine; and the same act which brings one

---

* A portable distilling apparatus of great productive capacity occupies an inconsiderable space in sailing ships.

into operation, throws the other out. The superior merit, and the distinctive peculiarity of the "link motion" is, that with it, the same one act which reverses the eccentrics, also reverses the valve.

47. With the link motion, the eccentric rods do not hook and unhook as in Arts. 35 & 36, but they have permanent hold, with the eccentric pins, at opposite ends of a short iron arc, or *link* curved to correspond with an arc. The link plays back and forth on a pin in the head of the valve stem (r in the Fig. p. 56), so that this pin may be brought to a position in either end, or in the centre, or in any other intermediate point of the link, at pleasure.

48. When the link is moved so as to bring the valve stem pin into its center position, that pin then becomes a pivot, without any rectilinear reciprocating or other motion whatever, being entirely at rest; and the two ends of the link, in other words the two eccentric pins, when the shaft is revolved by hand (no steam being on), will vibrate in equal arcs, as opposite equal arms of a beam do on a centre. Consequently, the valve attached to the stem, as in the Fig. page 56, will likewise remain at rest, notwithstanding the hand revolution of the shaft, and the vibration it gives to both eccentric pins at the opposite extremities of the link. This point of equal vibrations, is the *still point*.

49. And if, when the valve stem pin is thus in the centre of the link, the valve has no motion, but is at rest in the centre of the steam chest, covering all the three passages seen in the Figure page 56, which shuts out the admission of steam to the cylinder by either passage n or s, it follows, that so long as the pin holds this central position in the link, the engine cannot move by steam; for however it may be in the chest, it cannot reach the cylinder to act on the piston.

50. But alter the link on the valve stem pin so as to bring the pin nearer to one extremity of the link, and then a revolution of the shaft by hand as before, will cause unequal vibratory motion of the opposite arms and eccentric pins, and a reciprocating rectilinear motion of the valve stem pin, consequently also of the valve, will take place. And the nearer the valve stem pin is brought to either end of the link or either eccentric pin, the greater will be the motion of the valve, the more the steam and exhaust passages will be opened, and the greater will be the action of steam when there is any in the steam chest to flow through the passages.

51. When the progressive eccentric (Art. 46) is that which is nearest the valve stem, the engine under steam will work ahead; and when the retrogressive eccentric is that nearest the stem, the engine will work back.

52. Therefore, steam being up, when the engine is at rest the link stands midway on the valve stem pin. To go ahead, move the link by the lever, or by the wheel, or by steam, applied to facilitate the movement, and bring the progressive eccentric nearest to the stem. To stop, move the link to bring the two eccentrics equidistant from the valve stem. To back, bring the retrogressive eccentric pin nearest to the valve stem. To work strong either way, move the link with one end or the other close up to the valve stem; and to go slow, ease the position of the pin back towards the centre of the link. When worked by the steam it reverses in 45 seconds; by leverage rarely in less than 5 minutes.

53. Accompanying the slide valve, and this method of using it with the link motion, there are also the injection, the damper, and the safety valve, to be handled as in Arts. 35 & 36.

54. Although one of the most simple, as it is most ingenious and useful of the appendages to the screw engine, it is nevertheless somewhat difficult for the mind of a person not accustomed to consider mechanical detail. But with this description in hand, and the instrument present, especially if in operation, it may, by a little patient study, become well and familiarly understood.

SECTION VII.

Construction of Steamers. Side wheels and Screws. Wheel Shafts. Screw Shaft. Elements of the Screw. Steerage. Conclusion.

CONSTRUCTION OF STEAMERS.

1. A prominent consideration in constructing steamers, is to obtain in them the least resistance proportional to the displacement, consistent with the strength and stability requisite for the service to be performed. If this service regards only speed under steam alone, and is to be performed in smooth water, the resistance may be reduced very much by giving great length as compared with the beam or breadth. In this manner the displacement may be doubled without an increase, but on the contrary a reduction of resistance, by rendering the water lines "easier;" that is, by reducing the angles with which the vessel enters and leaves the water. For smooth water, there is scarcely a limit to the application of this principle, except steering in crooked channels and turning in comparatively narrow places, also that imposed by the friction arising from length.

2. But these excessively long vessels are objection-

able as steamers on the ocean for several reasons. Of these, one is the enormous weight of the engines and boilers concentrated within a small space near the centre of the vessel, which, when the two extremities are sustained by the tops of two waves, being partially forsaken by the trough of the sea, will settle, and occasion leaks, unless the vessel is constructed with an extraordinary degree of strength proportioned to the length.

Again, if a very long vessel, heading a heavy sea, is raised at the bow by a wave, and that wave passes under her to the centre, sustaining that part, the bow will overhang the wave and drop, opening the butts of the planks, and occasioning there also strain and leak.

3. The kind and degree of strength necessary to prevent the extremities, and the centre, of a long steamer, alternately settling in the manner described, are given chiefly by the side planks. If the sides are deep, so that this planking has great breadth, the vessel will be correspondingly strong;—otherwise weak. Several long river boats with no great depth of sides, have broken at sea and foundered.*

A limit to the strength produced by depth of sides is prescribed by the practicable height and depth of the vessel, which must bear a certain relation, and both may be too great; one for stability and as affording an object for opposing winds, and the other for draft of water and passage of bars found at the entrance of most harbors.

4. A second objection to these excessively long

---

* The "hog frame" is an expedient to compensate for want of depth of sides, but at sea is not reliable. The bow and stern too, not being "water borne," are hung by braces, and other expedients, which also, although well enough in smooth water alone, and well enough as auxiliary to deep sides at sea, are not there a good sole dependence.

vessels is, that if in a gale steam fails, they fall into the trough of the sea, and there remain in spite of every effort hitherto tried, sails or drags, and wallow until their decks are swept and they founder. The San Francisco and Central America are memorable instances.

A third objection is urged in certain cases, as men-of-war to compose the body of a fleet, which it is desirable to compact, and manœuvre quickly, and within a reasonable space. *

5. In regard to size of vessels, their capacity to carry fuel, power, &c., is as the displacement. The resistance, to which the power must be proportioned, is as the area of the greatest immersed section. But as vessels increase in dimensions, their forms being similar, the capacity increases as the cube of any given dimension; whilst the area of the immersed section, consequently the resistance, increases only as the square of that dimension.† Hence, increasing the size

---

* Long sailing ships have relatively an advantage in speed, pitch less, and are much more weatherly, because the lateral resistance is greater proportionally to the longitudinal, and because they brace the yards sharper; but they won't stay so surely unless the head yards are checked in, because they lose headway before the yards braced extra sharp catch aback. They cannot either be got off the wind in a squall, therefore need more careful watching. They require an inconvenient space staying, and more for wearing—an inconvenience especially felt in fleets.

The English complain bitterly of the unmanageable character of their new long Steam Frigates copied after ours, which is attributed to length. Their long rows of battery on a single deck, are ridiculed as " streets of guns." In truth they would, in line, fare badly against the concentrated fire of a two or three decker;

nd will, if so be it turns out by actual war experience that the line system is to continue. This, however, by the best opinions, will not prove the case. If it does not, and the melée system prevails, then ships fighting under steam, will as often be engaged on both sides as on one, obliging them to fight both batteries, each with half a crew, as rapidly as one battery can be fought with a whole crew. As guns are now mounted, this would be impossible. The Author is prepared with a means of meeting this new necessity, and he will propose it in due time.

† Solid measures increase with the cubes, and superficial as the square of a linear measurement. Hence, whilst the space in the ship increases as the cubes, the surface on which the carpenter works, increases only as the square, which accounts for the reduced proportional cost of large ships; and it would be less than it is, except for the scaffolding and hoisting on the stocks.

of a vessel so as to double her resistance, and double the cost of running her by doubling the quantity of fuel consumed in a given time, *more* than doubles her capacity to carry freight, fuel, &c.; which explains why large vessels of any kind are found most profitable where there is employment enough for them, and why large steamers can keep the sea longer, and accomplish longer voyages with the fuel they are capable of carrying, than smaller steamers.

Let there be taken, for example, two vessels, one 30 feet wide, 150 feet long, and drawing 10 feet water; and another 40 feet wide, 200 long, and also drawing 10 feet. The displacement (or capacity to carry) of one is represented by 45,000, the product of the three measurements; and the displacement of the other by 96,000. The relative resistances are represented by 300 and 400.—That is, the capacity of the larger vessel is more than 100 per cent. greater than the smaller, and her resistance, and consequently her power and expense, are greater by only 33 per cent.

6. But, by Art. 3, the depth must grow with the length, at sea, for strength. It must also increase in order to give lateral hold in the water to correspond with the lateral exposure to the force of both wind and sea, for otherwise the drift is such, that however the ship may head, no one can know the actual position on the sailing chart, owing to this great and uncertain drift as a cause of deviation. Hence the gain of speed by length, although always great, is in practice reduced below the figures of the preceding paragraph.

7. In proportioning engines to vessels intended for steaming only, it is customary to allow a horse power for every one, two, or three tons—giving the highest proportion to smallest vessels, for reasons noted in Art. 5. There is a growing partiality for high pro-

portional power, especially for vessels engaged in the transportation of passengers, yet there is much argument as to what the limit of this proportion should be. A correct solution depends on the purpose of the ship, whether for a man-of-war or not. If it be speed, dispatch, packet service alone, sacrifice largely, every thing to speed; otherwise not. And so with such men-of-war as are built for speed, to run, or principally for that. Or if they are built principally to claw off a lee shore-as some strangely contend, then give them a power adapted to this main object of their construction, otherwise not. And if they are to perform service about home exclusively, they need one construction and proportion of steam power to tonnage; that it should be primary, not secondary or auxiliary to sails; otherwise the reverse. For fighting and for distant service, ships undoubtedly require battery, spars, and subsistence, which are opposed to excessive proportional steam power, or to the weight and space it occupies.

8. The law of power in its relation to speed is, that power increases or decreases with the cube of the speed; and calculating, the statement is, as the cube of a given speed, is to the power which by experiment produces that speed, so is the cube of any other required speed, greater or less than the given one, to the power which will produce that required speed.

Thus, if it is known that in a given case 500 horses power will produce a speed of 8 knots, and it is desired to know what the increase of power must be to increase the speed $\frac{1}{4}$, or to 10 knots, the statement will be, as $8^3 = 512$, is to 500 (H. P.), so is $10^3 = 1000$, to 976 (H. P.), or nearly double the power. So that doubling the power produces only $\frac{1}{4}$ increase of speed. By trying other cases it will be found, uniformly, that doubling the power gives about $\frac{1}{4}$ increase of speed.

Hence a moderate increase of speed involves an enormous increase of weight, and demands room correspondingly for engines and boilers, and more yet for coals.*

9. In regard to the water lines of vessels, experiments long ago determined, that the form of least resistance had its sharpest end forward. But short sailing vessels so built, buried, and have even run under and foundered. Long vessels are in no such danger. Nevertheless, it is but recently that constructors have boldly conformed practice to theory, and brought the dead flat amidships. Mr. Steers led in this step, and hence mainly his success. Steamers, which are such exclusively, are often much fullest aft. †

10. Ships, of course, freight around the weight of their hulls,‡ and it is desirable that so far as possible

---

* It is truly desirable that the public, which properly regards speed as the chief merit in packet and passage steamers, should regard men-of-war with more reasonable and charitable criticism, remembering they are designed for distant, long-continued cruising, away from supplies of fuel; and besides engines, must carry heavy batteries, heavy masts and spars, subsistence and water for large crews for many months—a lading wholly incompatible with the lean water lines, and the heavy boilers and engines which conduce to mere speed.

The proportions of horse power are given in the books as relating to tonnage, sometimes to displacement, and sometimes to area of immersed section; and in reading intelligently, it is necessary to know which is meant: neither being expressed.

So also there is, beside the calculated and the indicated horse power already explained, another one spoken of in English books, termed the "nominal horse power," and in reading intelligently, it is necessary to know also which of them is meant, when neither is expressed.

Nominal horse power, as used in English publications, expresses the relative capacities of cylinders, and the work the engine will do with some certain effective pressure upon the piston per square inch, the books say 7 lbs. (Bourne p. 50); but is no measure absolutely of the work an engine does.

† Large ships with *short floors* invariably fail at sea, though fit for smooth water.

‡ On this principle, of the impossibility of freighting all around the globe any number of vessels loaded with their hulls, the "coat of mail ships" now bugbearing the world, will prove wholly impracticable as cruisers, although for special service against a neighboring belligerent power, they may no doubt prove effective, more particularly if ever it turns out that they are made impervious to heavy shot.

So also the "Steam Ram," which must be of enormous weight and strength, although of some service about home (yet even then far short of what its cost should render it) may very likely turn out a *sheepish* affair. Certainly it should be permitted to sink but one vessel, and that one should take the Ram down "by

each part of the ship should carry its own weight. This the bows and sterns of very long sharp ships do not; in other words, those parts are not water borne, but are as much hung to the body, as a horse's neck and head, and are to be held up by a heavy and expensive constant support. This very difficulty imposes another check upon length, and still more upon sharpness; for art must yield to nature—planks and bolts to gravity.*

11. It is useless to complain of the expense of a steam Navy, for there is no avoiding the greater first cost of ships, the more frequent repairs arising from the shake of the engines and the rapid decay caused by heat, or the larger amount required for pay. The Wabash, after but two years' service, shows in her wales, midway of the ship, only a shell one inch thick of sound wood, although at and towards the extremities, away from the heat, the planks are good the whole thickness. This may in part be due to unseasoned stuff used in the hurry of building, for undoubtedly steamers require the very best of seasoned material—at least in the middle, or waist.

SIDE WHEELS AND SCREWS.

12. The side wheel, is to the screw under steam power, what the paddle—more properly a pair of

the horns," head foremast. In war, defence always keeps pace with the attack, and following the Ram's introduction, will be appliances for grappling it on the instant, if not before all'the fatal damage is effected, yet before the victim can sink, so that when the Ram takes that projected "turn back," it will find " its horns caught in a thicket "—that it is easier to get into a scrape than out of it. Will the Rams carry their extremities in a heavy sea, or will the steel-plated ships carry theirs as cruisers ?

* The mania for increasing length will hardly be cured, until after more disaster. But unfortunately the victims will be a simple public which knows no better, intent only on going ahead, and not the capitalist and architect who don't go to sea in the vessels, only order and construct them, under the united impulse of cupidity and vanity.

paddles, or banks of oars, are to the scull under hand power. And the parallel only fails, because so much hand power cannot be brought to bear on the scull as on oars, whereas an equal steam power can be brought to the screw as to side wheels.

13. Even if the parallel did not fail for the reason mentioned in the case of hand power, and so much hand power *could* be brought on the scull as on oars, relying alone on the "ash breese," a figurative term for the oar, they would be voted preferable to the scull in smooth water; although in rough water, or co-operating with sails, all experience demonstrates the imperfect, awkward action of oars.

14. Throughout nature, where motion alone is the object, the rotatory is that which is always witnessed; and in art, where motion alone is the purpose, nature is imitated with analogous benefit. Under such circumstances, then, there is an advantage in bringing the rotation of the crank shaft to act directly as propulsion by the paddle boards or buckets, rather than indirectly and obliquely by the screw.

15. The screw, therefore, like all intermediaries, like for example the gearing Art. 17 page 62, may be regarded as a necessity, introduced to avoid some difficulty otherwise unavoidable, or to gain some advantage otherwise unattainable; the particular difficulty in this case to be avoided being the unequal action of side wheels in rough water; and the particular advantage sought being a union of the elastic force of steam produced by artificial means, with the natural force of the winds on sails, which is a result of gravity. Art. 1, page 9.

16. None would think of any other appliance for speed on a railroad, than the driving wheel acting directly by traction. Only where traction is insufficient,

it has been proposed to overcome inclined planes by a screw. So afloat, on smooth rivers, where an even keel and even action of the paddles is always possible, the case is very near akin to that of railroads. Hence on rivers, side wheels are usually seen—screws never. True, a lack of depth or draft of water to submerge a screw, is an additional reason for its absence from rivers; but without that reason, it yet wouldn't be there.

17. Early experimenters in this country, those coeval with Evans, Fitch, Stevens, and Fulton, essayed with the screw, and developed its advantages in deep water with a sufficient draft. But in shoal water, it could not be used even if desirable; and in smooth water it was not desirable. Hence the side wheel got the ascendency in America, where shoal smooth rivers and bays were the field;—an ascendency which doubtless the screw would have got instead, in England, where the boisterous channels and their deep water were the field demanding steam power to navigate them. Naturally, in copying from us who led in steam navigation, the English took the side wheel, which was also best adapted to the Boulton and Watt's form and style of engine, then universal; and although the screw proves now to be best adapted for channel service, it is not wonderful that time alone could break the hold which possession gave upon prejudice for the side wheel, as it has now done there, and also begotten a new form of engine, the Screw Engine, adapted to the work required. Nor is it wonderful that we are behind England in screw propulsion, and even for ocean navigation reluctantly abandon the side wheel, originating with, and handed down to us by an ancestry whose memory we venerate, and whose genius and perseverance merit our own, and challenge the world's admiration.

18. Side wheels, to operate with only small loss of power consequent on the buckets or paddle boards (when fixed to the arms of the wheel) entering and leaving the water at an angle with its surface, have very great diameter; an evil of which is, that it causes lofty wheel houses, and great retardation from head winds, as well as injury to the stability of a vessel.

19. The English very generally escape this evil of retardation and instability, by smaller side wheels, with swiveled buckets or paddle boards, so turned, by a "feathering wheel" on the shaft, as to preserve them always in a vertical position. Hence they enter and leave the water vertically, however great the dip of the wheel; whereas, the fixed buckets ("floats"), even of a larger wheel, increase or decrease their angle of entrance, to some extent, as the dip increases or decreases;—which dip is, of course, at the beginning of a long passage, very great, and at the end very light.

20. So also when the lee wheel of a side wheel sea steamer under sail is buried greatly, a similar action takes place; that is, a great loss of power, by the fixed buckets entering and leaving the water with an action which, to the extent it is vertical, is not propulsive, therefore lost; and which, if a wheel were buried to the shaft, would be *wholly* vertical. With the swiveled (the English call it the "feathering," as distinguished from the fixed, which they call the "radial") paddle, what force that paddle does exert, even in the extreme case supposed, is horizontal, and in no degree vertical.

Under canvas, the weather wheel dips lightly in proportion as the other dips deeply, and it is then of little account whether the paddles of the weather wheel are "radial" or "feathered." Under great heel, therefore, with side wheels there is great loss of power; and under any heel, the loss is proportional.

21. But there is another evil with side wheels, viz: back water action of all the paddles, whether "feathered" or "radial," attached to paddle arms which enter or leave the water at any considerable angles of obliquity. And this evil is greatest with small wheels. In fact, but for "slip of the wheel," which is the difference between speed of wheel and speed of vessel, and usually about 20 per cent. or $\frac{1}{5}$, every paddle except that on the vertical arm would be inoperative, or else back water. Any one arm entering or leaving the water at 45° or more, may be reckoned surely to carry a back water paddle; and probably those entering with a less angle. When a vessel by rolling, or heeling under sail, immerses a wheel more or less, but to a varying extent, there is constantly a loss of power in accommodating speed to this back water.*

22. Therefore, whilst in one respect the large side wheel with fixed or radial paddles is best, and in another respect the small wheel with swiveled or feathered paddles; it may unhesitatingly be declared, that neither of them is, in any respect, proper or fit for use as a means of propulsion in a sea way, or in conjunction with sails, or for a voyage—the draft of water in the beginning and in the end of which must be greatly different; in short, for ocean navigation.

23. The screw is altogether free from influence by the more or less deeply laden state of a vessel, by

---

* There is an analogy between this back water action of a paddle, and the cycloidal motion of any given point on a wheel rolling over the ground; and an explanation on that principle is often given. But there is a simpler one, and it is useless ever to go deeper in the well of science, than is necessary to find all the explanation a case requires.

Resolve the oblique motion of a paddle where it strikes the water, into its vertical and horizontal components, and if the horizontal is less than the speed of the ship through the water, there would be a back-water action, but for the slip.

heeling under canvas, or by rough seas, especially when in vessels of 15 feet draft and upwards. With less draft, sometimes the pitching motion is such as to throw a two-bladed screw wholly or in great part out of water, and occasion not only some loss of steam, but a dangerous and irregular speed of the engine. Devices for the spontaneous correction of this difficulty, peculiar to a screw vessel of light draft, are proposed. All of them act on the principle of the "governor." See note page 76.

WHEEL SHAFTS.

24. In shafting, several precautions are necessarily observed, as important; and that most so, is against damage from working of the upper frame of the vessel, and unequal settling of parts, particularly the wheel guards.

25. Each one of the side wheels has its separate shaft, with a main bearing at each end; the outer one on a heavy timber which spans from the extremities of the two guard beams, and the inner one on a crank frame erected from the floor of the vessel; or when there is but one engine, this crank frame is built up from the kelson. Both these bearings, by which the vessel is at last driven, are well braced forward and aft.* The shafts being of wrought iron (forged under steam trip hammers), each has a crank arm "shrunk on" to its inner extremity, and the connecting rod of the engine is strapped to a short "crank pin" between them, reaching from one crank arm to the other. But this crank pin, which is a firm fixture to one of the crank

---

* Each wheel shaft has also a spring bearing at the vessel's side, but it is not arranged to support the middle of the shaft when the extremity settle. It has though, firm braces both forward and abaft it.

arms, is neither keyed nor in any way immovably secured to the other; because, if opposite guards settle, it will occasion the two crank arms to spread apart, which they must be free to do without occasioning strain or fracture. This necessary play is given, by what is called a "drag link," which any person ought by inspection readily to comprehend the use of.

When there are two engines, an intermediate shaft is put in between the starboard and port crank frames; and each extremity of this intermediate shaft carries a crank arm, which is provided with the drag link.

THE SCREW SHAFT.

26. The screw is either attached, or fixed to a longitudinal shaft, extending from just abaft the engine (placed usually in men-of-war just abaft the mainmast, which steps between the engine and boiler*), along the shaft alley, over the kelson, to the stern, where it passes out by an orifice bored through the dead wood, and in case of a lifting screw, through the main stern post. The shaft has a principal main bearing in the stern, and another principal main bearing at the other extremity near the engine; where it has also a circular

---

* Nothing in the economy of a steam man-of-war's arrangements, has been more considered, or given rise to a greater variety of practice, than to step the mainmast so as to bring the step, where it belongs, down on the kelson, and not on the berth deck, or on a gallows frame over the engine, or the screw shaft, or to straddle them; to permit the centre of gravity of the boilers and engines as a whole to lie near the centre of gravity of the ship, and at the same time to have no considerable loss of space between the boilers and engines, and give likewise no unnecessary length to the main steam pipe, which by length is more exposed to damage by shot, and to condensation of steam passing through it from the engine to the boiler; to throw the smoke pipe so far forward that it will not interfere with boarding the main tack on a wind, and yet leave the usual place for stowing the launch free for that purpose. These are the considerations to be reconciled, and it is a capital field for an officer's study and the exercise of his ingenuity, as well as a point for observation in the inspection of men-of-war, as they are met with, belonging to various nations.

clutch piece, corresponding with and fitting loosely to another clutch piece on the after extremity of a crank shaft, to which the engines connect. When the crank shaft revolves, it communicates motion to the screw shaft by means of the clutch.

The crank shaft is usually forged all in one piece, having two cranks set at right angles to each other, so that when one engine is on the centre or dead point, the other is at the half stroke; the effect of which relative disposition of the two cranks is, that one engine assists the other over the dead point, and evenness of motion throughout a revolution is maintained. These cranks, like all others, are carefully counterbalanced.

27. The clutch, by its two pieces not fitting closely, allows for the " hoging " of the ship, that is settling of the stern and with it the after end of the shaft, without a strain ; in which respect it accomplishes the purpose of a drag link to the side wheel crank. The screw shaft, being very long, is forged in several pieces, never exceeding 15 feet, and there is a main bearing where the lengths join, also an adjustable spring bearing under the middle of each length.

28. In case of the side wheel shafts, there are four main bearings to sustain the weight, besides the two spring bearings on the sides, and the force of the paddles results horizontally upon these several bearings, to drive the ship. But this force on the shaft being divided among the whole six bearings, that exerted on any one of them is not great. But with the screw shaft it is different. The whole propelling force of the screw, by which it acts on the ship to drive her, which force is termed the "thrust," must be exerted either against the stern post or frame, where lubrication would be impossible and the parts soon wear out: or endwise on the shaft to drive it in. either

against the clutch, or against some other obstruction placed expressly to receive the "thrust." Accordingly, every screw shaft has what is called a "thrust bearing," which is a collar arrangement on the shaft, crowding horizontally forward or back against a heavy timber framed into the ship. This also is easiest understood by inspection, and the aid of such oral explanation as may generally be obtained. The thrust bearing is away aft in the shaft alley, near the stern.

29. But the most important feature in connection with the screw shaft, that which has been found most difficult to perfect, and until perfected was the great want standing in the way of success to the screw as a certain and safe means of propulsion in heavy ships, is the stern bearing for the screw shaft, in the orifice through which it protrudes to couple with the screw. Whilst this was an ordinary metal bearing, it could never be made to stand, because of the enormous weight of the screw and shaft resting on it, the great rapidity of the revolutions, and its inaccessibility for lubrication. In some instances on board heavy ships, the bearings have worn away and settled, not only to produce obstruction, but to admit water, so as to endanger ships, and make it necessary to beach them to prevent foundering. An effectual remedy has been found, strange as it may appear, in wooden, lignum vitæ bearings, or metal cases lined with that wood. This, and a small flow of water in channels left between the wooden lining pieces, to keep down the heat arising from friction, now answers the purpose, as nothing else does; and almost every case of an attempt to dispense with this wooden appliance, has resulted in at least an impaired efficiency.*

* When working hawsers from the stern of a screw ship, be ever vigilant against their fouling the screw. See page 6.

COUPLING AND LIFTING SCREWS

30. The first screw brought into use at sea was Ericsson's, and the "Princeton" its first grand exemplification. Her performances were very creditable and successful, she having proved herself a most efficient man-of-war, especially by her promptness as a blockading ship at Vera Cruz. The British Admiralty tried it in the "Amphion," and the French Marine in the "Pomone" Frigate. For some reason, none of these experiments were repeated; Ericsson's screw went out of use at sea, and another one has taken its place—the inventor being an English farmer, Mr. F. P. Smith.*

31. Ericsson's screw *hung* by the shaft, and the enormous weight was sustained solely by the rigidity of the shaft, which needed to be correspondingly strong. When Fulton first applied side wheels to river boats, his wheel was hung in the same manner, by the shaft, with no outer or guard support. His greatest and long-continued difficulty, arose from inability to hang the wheel in this way securely. A workman is said to have suggested the guard support. Fulton's genius seized and adopted the suggestion, and success was immediate. Fulton's error, therefore, was Ericsson's. The distinctive characteristic, then, of Smith's screw, as compared with Ericsson's is, that the former has an outer support, or is at least steadied by an outer spring bearing, on the outer or after stern post to which the rudder is hung. And in searching for the reasons why Smith has been successful whilst Ericsson was not, it is probably to be found in the fact of this outer support. The only heavy screw ship now performing service at sea without an outer bear-

---

* See an able article on Screw Propulsion in the Atlantic Monthly, from the pen of Commander Walker, U. S. N.

ing, either as a main bearing for support, or a spring bearing to steady the screw, is the "San Jacinto"—and she has never been a reliable vessel *with her screw* on foreign service—although *with her battery*, gallantly commanded in China, she has performed most excellent and effective service.*

32. When the outer stern bearing is a main bearing, the outer stern post to which the rudder is hung, needs to be strong and large, which renders it a heavy drag, retarding in its effects, and causing considerable loss of power. But when the outer bearing is only a spring bearing to steady the shaft, the outer stern post needs less strength, is a less drag, may be and often is of metal, and thus occasions a very diminished or inconsiderable loss of power or speed.

33. In passage vessels or mail packets, in which steam is the principal power, sails merely auxiliary; which never uncouple to run under sail alone, and can afford neither the loss of power nor of speed produced by the heavy stern post; the outer bearing is invariably a spring bearing to steady the shaft, and the post is of metal, producing very small resistance or drag. And when for reasons extraordinary, such as accident to the machinery, it becomes necessary for these mail packets to uncouple, so that the screw may revolve freely, the uncoupling gear is found forward of the "collar bearing" provided to receive "the thrust," (Art. 28); by which the outer bearing still remains only a spring bearing, and the support of the screw

---

* Allusion is here made to an unacknowledged and unrequited service, performed chiefly by Commanders Foote and Bell, U. S. N., in capturing and destroying the "Barrier Forts," China, in 1856, and by it preparing the way for a most successful diplomacy.

By great care, and unusual skill, the ship was got through her China cruise; but her antecedents had not been, nor is her subsequent history, calculated to engender confidence.

continues to depend in part on the rigidity of the shaft, (Art. 31.)

34. But a man-of-war, on foreign service, relies on sails principally, carrying steam as an auxiliary, and must cruise a large portion of the time wholly or in part under sail, using steam only in emergencies, which may or may not be frequent. Her screw bearings are accordingly adapted to this peculiar necessity. Thus far, this adaptation seems to require, that the outer bearing should be, equally with the inner one, a main bearing; the outer or rudder post consequently a heavy one; and the drag and loss it occasions be submitted to as an unavoidable necessity. And when both stern post bearings *are* main bearings, the "screw axle" is made no part of the shaft, but rests with its two axle arms, one in each stern post, and may revolve independently of the shaft, or any part of the shaft, as it does when disconnected or uncoupled.

35. For a screw thus capable of a revolution on its axle independently of the shaft, the coupling arrangement is effected by protruding an arm (from within the after end of the shaft as from a sleeve), which enters the screw axle, that being a hollow cylinder fitted to receive the protruding arm, and in a manner, by means of a slot, to cause the screw to revolve when the shaft is turned by the engine. Such is the plan in use on board the English ships first equipped with Mr. Smith's screw, and adapted to the peculiar requirements of military service, as cruisers abroad.

36. A more recent improvement, universally applied to ships of war lately constructed, is "the well," in which, when under sail alone, the screw is hoisted entirely out of water, in lieu of coupling by means of the arm protruding from the shaft as a sleeve, described in Art. 36: and the screw axle is solid instead of hol-

low. The details of the mode in which the screw is thus alternately hoisted, and lowered again into coupling with the shaft, so as to revolve with it, are best learned from observation, inspection, and enquiry. It is a most ingenious arrangement, due, it is said, to a French officer, and obviates a difficulty, viz.: that although when, with high speed of the ship under sail, an uncoupled screw left in the water free for revolution will so revolve and produce very little retardation, with a speed of only 4 or 5 knots the screw does not turn but is wholly a drag. So when with high velocity it does turn, the jar, noise, and wear produced, are worth obviating, and *are* obviated by lifting the screw out of water.

37. Another reason of governing force, yet not always considered, why the outer stern post for the bearing of a screw axle which may revolve independently of the shaft or any portion of the shaft, must be heavy and strong, when the stern post for a screw which is fixed to the shaft need not be, is, that if the former screw is turned back strong by the engine, the entire backward thrust results on the stern post, which, if light, would give way; whereas in the latter case, the uncoupling being effected forward of the thrust bearing, Art. 34, so that the after part of the shaft revolves with the screw, a "collar thrust" bearing on the shaft is so contrived (Art. 28), that it receives the backward as well as the forward thrust, and entirely relieves the stern post from that necessity for strength.

ELEMENTS OF THE SCREW.

38. The elements of efficiency in a screw, to be considered in comparing one with another, relate to revo-

lutions, to pitch, to diameter, and to the number, shape and surface of the blades.

With side wheels, the revolutions being alike, speed of vessel is as the diameter of wheel. With screws, revolutions being alike, speed is as the pitch of the screw, and has no relation to diameter, except that it gives surface; and if the diameter be less than is adapted to a vessel of 13 feet draft, the screw has not sufficient submersion to give it a proper hold in the water, and prevent an inordinate "slip"—slip being, in case of a screw, the difference between speed per log, and that due to pitch multiplied by revolutions. It varies from 10 per cent. under the most favorable circumstances in smooth water, to 20 ordinarily; and when a vessel can only stem a gale, the slip is 100 per cent.

40. By "pitch" is understood, such an inclination of the blades to the water, as will, in an entire revolution (the slip not considered) give any certain progress to a vessel—screw her ahead, and is reckoned in feet. Thus the Princeton's screw had a pitch, the highest recorded, of 35 feet. With a turn, then, slip not considered, her progression should have been 35 feet; with 20 per cent. off for usual slip, 28 feet. Her revolutions were 36 per minute. Therefore, $28 \times 36 \times 60 = 60,480$ feet per hour, or less than 10 knots (there are $6086\frac{7}{10}$ feet in a sea mile) per hour, should have been her speed. *At sea* in rough water, she never, however, did hardly 9 knots, which shows the slip there to have been greater. In all cases, it increases with the resistance of wind and sea, until, as remarked in the preceding article, when a vessel can barely stem the weather, the slip becomes 100 per cent., like when fast to a wharf.

41. The usual pitch is 18 or 20 feet. Sometimes it is uniform—a "true screw;" at others, the pitch is

increasing towards the extremity of the blade; which increase of pitch is with the same object as the "wave bow" (concave bow water line) of a ship, viz., more quickly to follow up the receding water. "Bourne," page 107 says, "the uniform pitch is as good as any," and "that no advantage has been found to result from an increasing pitch." He further recommends, "as large a diameter as possible, a quick turn, and a fine pitch."*

42. A steeper pitch is best for carrying sail, because a fine pitch increases the revolutions more under high velocities from winds and sails, and is most likely to occasion drag of the screw. Drag is easily detected, by multiplying pitch into revolutions per minute, and again by 60, then dividing the product by 6086 (the feet in a sea mile or "knot"). If this quotient is less than the speed per log, the drag is sure.†

43. As regards the number of blades, Bourne says, "a screw of two arms, or a portion of a double threaded screw, has been found as effectual a propeller as any other; but a screw of three blades, or a portion of a

---

* To a seaman's eye, the blade of a screw appears to have constantly a decreasing pitch towards the extremities of the blades, when in reality, and to a mechanic's eye, the pitch is not decreasing, but uniform.

A screw, in scientific mechanics, is but a form of inclined plane. Erect a perpendicular equal to half the pitch of a screw in feet; establish points on the base, at distances from the perpendicular successively equal to twice the distances of any assumed points on the blades from the centre of the screw axle; draw hypothenuses successively to the several points so established on the base; and these hypothenuses, by their decreasing angle at the base, whilst the perpendicular or half pitch which it represents remains constant, will indicate the decreasing inclination of the blades to the water towards their extremities:—in other words, that which appears in the screw, *is* a decreasing angle of inclination, but *not* a decreasing pitch.

† A screw, known as Griffith's, has been used, one characteristic of which is, that the pitch is adjustable, and can be increased—rendered steeper, which avoids an increase of revolutions when under canvas with good winds. But considering the immense force—a pair of engines, acting on only two arms of one propeller, it must be doubtful if they do not need the strength which belongs to permanence.

Side wheel engines divide their force between the two wheels, and again amongst several floats of each wheel; and because *they* admit of feathering (Art 19, p. 103), it by no means follows that screw blades will.

three threaded screw, has been found to act with a more equable and regular motion." In light draft vessels it is most important to have three, because in pitching, two blades may both be out of water at the same time, causing the engine to act with no resistance, and with dangerous rapidity. Three blades are, however, incompatible with the "well."

44. The area of screw surface is as the number, width, and length of the blades. And as the slip of a wheel decreases with the increase of the bucket, float, or paddle board surface, so ought slip to decrease with the increase in area of the screw—the screw being supposed constantly submerged.

Bourne says, "the length of screw that is found most beneficial, is about $\frac{1}{6}$ of a convolution;" by which he is supposed to mean, that the screw surface should be that produced by such width of blade—the width increasing, from the hub out, with the length of blade; which increase of width also preserves the relation of $\frac{1}{6}$ at all points with the "convolution."

45. But the best shape of blade is undetermined, for some are seen broadest in the middle (as Griffith's for easier "clearance"), others near the screw centre, others again enlarge uniformly to the extremities.

46. Sir Howard Douglas in his "Naval Warfare with Steam," page 61, proposes, with a view to reduce the "shake," to curve the leading edge of a blade, so that it shall not enter or leave the water all at once, but gradually; and moreover, that these leading edges should, for men-of-war, be made sharp, to cut or saw obstructions threatening to choke or impede the screw. In battle, the screws of those vessels are peculiarly exposed to disability, by spars, shot away and floating about, and the rigging hanging beneath the surface from them. Sir Howard's plan of a curved blade

edge, has great apparent merit, and is said to have accomplished the very important purposes intended.*

STEERAGE.

47. Sir Howard justly remarks, (ibid, page 72), that the steering of a screw ship-of-war, particularly when manœuvring under steam alone, "should be as if instinct with life, intuitive, quick as volition!"

These screw vessels do steer better, quicker, and turn in much less space under steam, than side wheel ships; and for the reason, that the currents thrown by slip of the screw against the rudder, counteract the dead water which proverbially impairs its efficient action; whereas the side wheel, by its slip, produces currents which give an apparently increasing speed of vessel through the water, and cause at the stern a corresponding actual increase of dead water. Hence side wheel steamers require, and are found to have most rudder, in proportion to tonnage, length, and displacement.

48. When a ship is under sail alone, or with a tow, and the screw is coupled but drags, or is uncoupled, and the rate of sailing so slow as not to revolve it, especially if there are but two blades and they set in the vertical position, the inclination of the lower blade will act on the steerage like a rudder with its helm over to one side, because the upper blade, although inclining equally the other way, does not produce en-

---

* Instead of wasting power by crowding the screw through a narrow space between stern posts set near to make a narrow well, which is a greatest cause of "shake," we save the power and in a measure avoid the shake, by a wider space and larger well than others use. We avoid also a sacrifice of screw surface where it is most effective, viz., at the extremities of the blades, whilst Griffith's blade obtains clearance (Art. 45) by this sacrifice. These considerations are thrown out to engage the attention of seamen, and direct their observation; for the seaman and the engineer are very necessary coadjutors. On *fouling*, see pp. 6 and 108.

tire neutralization, but it has to be produced by au opposite action given to a rudder with the helm; and even that may prove insufficient. Hence a ship under these circumstances, will turn quicker, and in a shorter space, the way in which the lower blade and the rudder act in conjunction.

49. Again, when the ship is moving by the screw under steam, she will be found to turn to port in obedience to a starboard helm, more readily and in less space than she will turn to starboard in obedience to a port helm; and to keep a course by compass, it may be found necessary to carry a small port helm; the supposition being, that, looking forward, the screw turns *with* the sun (from left to right), as it usually does, and naturally should in screwing a ship ahead, otherwise it would be a left-handed screw. This effect upon steerage is caused by action of the lower blade revolving against a greater resistance than the upper blade meets moving in an opposite direction; and these opposite effects differ most in light draft vessels, where the upper blade is not always constantly and entirely submerged.

From ignorance or disregard of this peculiarity in screw ships, most disastrous collisions * have occurred.

---

* Though not relating to the present discussion, it is well to say, that for the purpose of preventing collisions at night, an order from the Navy Department requires Government vessels, when under steam, to carry three lights—a white light at the foremast head, a green light on the starboard side, and a red light on the port side. These colored lights are screened and mutually seen only by vessels meeting. A vessel therefore seeing, for example, a stranger's red light only, in the direction its own red light shines, knows that the stranger and itself are on nearly opposite courses, with no danger of collision; that if it meets a green light only, in the direction its own red light shines, the stranger and itself are on courses angling to each other, and if his bearing does not change there is danger of collision, otherwise not; and if both colored lights of the stranger are seen right ahead, both vessels should immediately change their course so as mutually to exhibit the red light only, by which each passes on the other's port hand. Generally, when vessels see from each other one colored light only, and that of the same color, they are safe. Where there is doubt about the bearings in case opposite colors are those mutually visible, the safest solution is for each at once to exhibit its red light

In time, however, seamen will become familiar with it, and learn instinctively to make the necessary allowance. *

CONCLUSION.

The foregoing pages contain all, it is believed, both of construction and practice, important to be known by any one not perfecting himself as a professed engineer; enough for the special necessities of the seaman; enough also for the general reader, deriving daily advantage from steam, yet exposed in a corresponding degree to its dangers. The popular mind is blissfully ignorant of steam, except as instructed by the chapter of horrors periodically revealed. Yet there is no folly in obtaining from more harmless sources, that degree of wisdom which will constitute the public a judge and a check over engineers, at sea and ashore, as it now habitually is over the other professions:—a corrective greatly needed by the times, and one infinitely more effectual than legislation!

only, which is in all cases equivalent to "keeping to the right as the law directs;' or if that involves an inconvenience, the next safest plan is to mutually exhibit the green light only. To make all this clear and familiar, sketch and study diagrams of all conceivable relative positions. There is, amongst Governments, a conventional understanding on the subject.

So, to avoid collisions a system of bells is established, by regulation or custom, for communicating speedily from the deck to the engine room of a marine steamer.

The Navy Regulation is: Ahead slow, 1 bell; fast, 4; slow again, 1; slower, 1; stop, 2; back, 3. The custom generally prevailing in the Merchant Marine is, ahead slow 1; fast, 3; slow again, 1; stop, 1; back, 2. Either is good. But if one is best, it ought to prevail; for uniformity is the surest guard against mistakes. The first is most complex, but least ambiguous.

* It is said that in calm smooth weather, by alternately throwing a current from the screw against a starboard helm, then reversing the screw to stop headway, a ship can be turned to head in an opposite direction without moving more than ner length. So at anchor, by throwing a current against a starboard helm, more properly against a rudder in the position which a starboard helm gives it, the direction of a broadside is in some measure under control of the helm, and so far obviates the necessity of a spring on the cable. Try it.

A screw does not back so effectually as a side wheel, because the water thrown forward has no free escape, but strikes the ship and reacts upon the screw.

INDEX.

A.

Adjustable screws, 114. Do. cutt offs, 114. Air pump, 53-55. Amphion, 109. Anthracite, 109. Argand burner, 51. Astral lamp, 51. Atmospheric weight or pressure, 15, 16. Armor, sub-marine, 6.

B.

Barometer, 15. Back action engines, 61. Banking fires, 90. Back water (paddles), 104. Bell, 110. Bells, 118. Bearings, main, 111; spring, 110; metal, 108; wooden, 108; collar, 110. Beam engine, 59. Bed plate, 55. Bituminous coal, 48. Bilge injection, 57. Do. pump, 57. Blast, 47. Blades (screw), 114, 116. Blow cocks, 26. Boilers, material, 45; bracing, 44; water and steam room, 44; form, proportions, and position, 36-46. Bridge wall, 51.

C.

Caloric, latent and sensible, 31-34. Caloric engine, 68. Channel to bed plate, 55. Chimney, 45, 47. Circulation of steam, 58. Cistern, 57. Circulation in water, 13, 24. Clearance, 80. Clutch, 107. Colored lights, 117. Collar bearings, 110. Coupling, 109, 111. Comparison in consumption of fuel, 82; of high and low pressure, 73; of engines, 62; of boilers, 59-64; of screws, 113; of side wheels and screws, 100; of coals, 49. Conversion of condensing to non-condensing, 73. Counterbalance, 76, 107. Connecting rod, 58. Condensing engines, 52-55. Combustion, 46. Construction of ships, 94. Conduction of heat, 13. Condensation, 11-12. Cocks, gauge, 23; feed, 66; injection, 57. Crank and pin, 58. Cutting off steam, 72. Cylinder, 52, 57.

D.

Decay of steam ships, 100. Delivering of waste water, 55. Diameter of screw, 113. Do. of wheels, 113. Direct action engines, 60. Distillation, 90. Donkey feed pumps, 86. Driving wheels, 77. Draft of furnace, 37-47. Drag link, 106. Drag of screw, 114. Duties, relations in, of Commanders and others with engineers, 8.

E.

Ebullition, 13, 15, 22. Effective pressure, 54. Exhaust, 55. Elasticity, 9, 10. Elements of the screw, 112. Engines, 61, 62. Ericsson's caloric engine, 68. Ericsson's screw, 109. Evaporation, 11, 30. Evans, 73, 102. Explosions, 22, 25, 89. Expansion of steam, 66, 72.

F.

Feed pipes, 57. Feed, temperature of, 66, 74. Feathering paddles, 103. Fire surface, 14, 37. Fire rooms, position and temperature of, 40-41. Fighting on a vacuum, 70. Fitch, 102. Floors, length of in ships, 99. Flues, 37; cleanliness of, 28, 83. Foaming, 21, 23. Foot valve, 55. Force pump, 57. Foul air pump, 68. Foote, 110. Fouling the screw, 6, 108, 115. Fuel, 48; economy in burning, 82. Furnaces, firing them, 82. Fulton, 102, 109.

G.

Gauges, glass, 24; cocks, 23; steam, 18; vacuum, 78; thermometer, 19. Geared engines, 61. Governor, 76, 105. Great Eastern, 83. Griffith's screw, 114, 116. Guns, author's plan of mounting, to fight both sides, 96.

H.

Hanging shafts, 109. Heat, sensible and latent, 31. Heaters for feed water, 66. Heat, specific, 35. Heating surface (see fire surface), 37. High and low steam compared, 34. High pressure engines, 93. Hot air engine, 68. Horse power, calculated and indicated, 77. Hoging, 107. Hog frames, 95.

J.—K.

Jacketing, 72. Injections, cocks and pipes, bottom, side, and bilge, 54–57. Indicator card, 79. Indicated horse power, 77. Inclined engines, 60. Increasing pitch, 114. Knot, length of, 113. Isherwood, 38, 45, 49.

L.—M.—N.—O.

Land engines, 75. Length of vessels, 96. Link motion, 91. Lifting screws, 109. Low steam, high steam expanded, 34. Long stroke of piston, 60. Mail packets, 110. Martin's boilers, 40. Marine economy, 82, 86. Mercurial steam gauge, 18. Do. vacuum gauge, 78. Motion, 58, 101. Negligence of engineers, 29. Oxygen, 46. Oscillating engine, 61.

P.—Q.—R.

Parallel motion, 58. Paddles, radial and feathering, 103. Perforation of furnace doors, 51. Piston, 52, 58. Pitch of screw, 113, 114. Plated ships, 100. Power, sources of, 9, 10; proportions of, 98. Position, relative of boilers, engines, and masts, 106. Pomone, 109. Pressure, effective, 16. Prime movers, 9, 10. Priming, 26. Princeton, 109, 113. Pyrites, 50. Radiation, 72, 84. Raising steam, 86, 87. Rams, 99. Radial paddles, 103. Reservoir, 55. Revolutions of engine, 62. Rock shaft, 58. Running on a vacuum, 69.

S.

Safety valve, 17. Salometer, 28. Saturated water, 27. Saturated steam, 25. Scale, 27, 28, 66. Screws, 100, 115. Do. Engines, 62. Do. Shaft, 106. Semi-Bituminous coal, 48. Shaft bearings, 106. Shake, 115, 116. Side wheels, 100. Sir Howard Douglass, 115. Signal bells, 118. Side pipes, 57. Side lever engines, 59. Signal lights, 117. Slip, 104, 113. Slip joints, 87. Slide valves, 56. Smith's screw, 109. Smoke, 50. Specific heat, 35. Speed, 84, 85. Spring bearings, 105. Spontaneous combustion, 50. Speed of piston, 62. Square engine, 60. Steam with sails, 85. Stern posts, 110. Steerage, 116. Steam engine, 52. Still point, 92. Stuffing box, 52. Steam pipes, 52. Do. chest, 54. Stop valve, 57. Stroke, length of, 60. Steam room, 42. Steeple engine, 61. Steam drum, 26. Do. chimney, 26. Steam jacket, 72. Steering, 6, 102. Superheated steam, 25. Surface condensation, 64. Sulphur in coals, 49. Super-salted water, 27. Supports to shafts, 107.

T.—V.—W.

Temperatures of steam, 19–21. Temperatures, boiling, 14. Throttle valve, 50. Thrust bearing, 108. Ton weight, 83. Trunk engine, 61. Traction, 101. True screw, 113. Tubular boilers, 40. Vapor, 12. Water level, 22. Do. gauge, 23. Valves, safety, 17; steam, 54; exhaust, 54; stop, 57; foot, 55; slide, 56. Vacuum, 54; running on a, 69; gauge, 78. Watt, 67, 73. Waste of fuel, 37. Water room, 42. Water bottoms, 42. Water jacket, 67. Wabash, 100. Walker, 109. Well, 112. Weight, 9, 10. Wheel shaft, 105. Volume of Steam, 29. Worthington (see Donkey), 86. Wyoming, 30, 80, 83.

www.ingramcontent.com/pod-product-compliance
Lightning Source LLC
Chambersburg PA
CBHW031634160426
43196CB00006B/415